Metamaterial and Frequency Selective Surface Assisted Antenna Design

From fundamentals to novel design approaches

Online at: https://doi.org/10.1088/978-0-7503-5422-6

IOP Series in Electromagnetics and Metamaterials

Series Editor
Akhlesh Lakhtakia, *Pennsylvania State University*

Series in Electromagnetics and Metamaterials
The Series on Electromagnetics and Metamaterials, published by IOP, is an innovative and authoritative source of information on a fundamental science that has been enabling a multitude of transformative technologies for two centuries and more. The electromagnetic spectrum extends from millihertz waves to microwaves to terahertz radiation to ultraviolet light and even soft x-rays. In each spectral regime, different classes of materials have different kinds of electromagnetic response characteristics. Much has been discovered and much has been technologically exploited, but even more remains to be discovered and even more remains to be put to use for diverse applications.

Electromagnetics is an ever more vibrant arena of techno-scientific research. This is amply exemplified by the huge current interest in metamaterials. By virtue of carefully designed and engineered morphology, metamaterials exhibit response characteristics that are either completely absent or muted in their constituent materials.

Each book in the series offers an extended essay on a foundational topic; an emerging topic; a currently hot topic; and/or a tool for metrology, design, and application. Ranging from 60 to 120 pages, books are written by internationally renowned experts who have been charged with making the content not only authoritative, but also easy to understand, thereby offering more synthesis and depth than a typical review article in a journal.

Illustrated in full color for both ebook and printed copies, these short books are easily searchable in the ebook format. The series is thus more modular and dynamic than traditional handbooks and more coherent than contributed volumes.

This series is edited by Akhlesh Lakhtakia, the Charles Godfrey Binder (Endowed) Professor of Engineering Science and Mechanics at the Pennsylvania State University. Initial topics targeted in the series include symmetries of Maxwell equations, homogenization of bianisotropic materials, metamaterials and metasurfaces, transformation optics, nanophotonics for medicine and biology, single photons, radiation sources, optical bolometry, magnetic resonance imaging, and detection and imaging of buried objects. Additional topic suggestions are welcomed and will be promptly considered and decided upon. As part of the IOP digital library, students and professors at purchased institutions will have unlimited access to the ebooks for classroom and research usage.

A full list of the titles published in this series can be found at: https://iopscience.iop.org/bookListInfo/iop-series-on-electromagnetics-and-metamaterials#series.

Metamaterial and Frequency Selective Surface Assisted Antenna Design

From fundamentals to novel design approaches

Ayan Chatterjee
Department of ECE, Institute of Engineering and Management, University of Engineering and Management, Kolkata, West Bengal, India

Snehasish Saha
Department of ECE, Academy of Technology, Hooghly, West Bengal, India
and
Department of Engineering and Technological Studies, University of Kalyani, Nadia, West Bengal, India

Sushanta Sarkar
Department of Engineering and Technological Studies, University of Kalyani, Nadia, West Bengal, India

Partha Pratim Sarkar
Department of Engineering and Technological Studies, University of Kalyani, Nadia, West Bengal, India

IOP Publishing, Bristol, UK

ISBN 978-0-7503-5422-6 (ebook)
ISBN 978-0-7503-5420-2 (print)
ISBN 978-0-7503-5423-3 (myPrint)
ISBN 978-0-7503-5421-9 (mobi)

DOI 10.1088/978-0-7503-5422-6

Version: 20250701

IOP ebooks

British Library Cataloguing-in-Publication Data: A catalogue record for this book is available from the British Library.

Published by IOP Publishing, wholly owned by The Institute of Physics, London

IOP Publishing, No.2 The Distillery, Glassfields, Avon Street, Bristol, BS2 0GR, UK

US Office: IOP Publishing, Inc., 190 North Independence Mall West, Suite 601, Philadelphia, PA 19106, USA

Contents

Preface

Metamaterials have become the new cutting-edge technology for the wave phenomenon. After the inception of the concept of metamaterials almost a century ago, the research in this field of materials science took place in the last few decades. This progress was contributed partially due to the availability of technology for further progress of the research work on material science and partially to the need for the new materials for electronics and photonics. It was also partly propelled by the need for finer material for high-speed communication. As we moved from words per minute to gigabits per second, the need for newer, faster stable electronics was felt. This need was initially fulfilled by the VLSI technology. But the development of VLSI also reached its limit where the need for nano-technology based materials was felt. As we are moving from gigabit per second to terabit per second, the mobility of the electron is limiting the development of the speed. Hence, a paradigm shift towards photonics becomes inevitable. The research into metamaterials also resulted in its application in fields other than electronics. Metamaterials are currently also used in the sound industry as absorbers, and in the chemical industry for sensing chemicals.

A common problem that is generally faced by academics and new researchers for cutting-edge technology is the unavailability of proper study materials organized in a way to the first glimpse of the new field. The knowledge in the field is spread out in terms of vast number of research papers. The existing research articles have been presented on a different note. This book tries to address those difficulties faced by the researchers. This book starts with an introductory discussion about the concept of metamaterials and its modern day applications. This shall give the reader a clear picture of the end products that the research in the field is trying to achieve. This book also introduces antenna and frequency selective surfaces (FSSs). Thus, the book aims to provide the knowledge of both the metamaterials and the antenna and FSSs in one place. This shall be particularly helpful for the new researchers who are about to start research work in antennas and FSSs based on metamaterials. As the reader moves from one chapter to another, they shall learn how previous researchers used the phenomenon of the metamaterial for advancing different aspects of the antenna and FSSs. The book presents a few selected research works in metamaterials for application with terahertz antennas & FSSs, graphene based antennas & FSSs, conformal antennas & FSSs, reconfigurable antenna & FSSs and its application in biomedical field. The selected research examples that are presented in this book are simple in design and have been presented in a lucid manner. This shall also help a first-time reader in this field. In the final part of the book, 3D metamaterial-based design has been discussed. As the technology is fast paced, an attempt has been made to project a future path for this field of metamaterial research relating to electromagnetics. In the field of technology development, the future possibilities are infinite. The future projections drawn in this book are only a few among the infinite possibilities.

Although the book is intended for a new researcher, it shall be equally helpful for postgraduate students or anyone with a basic knowledge in electromagnetism/ antennas who just wants to acquire knowledge in the field. This book is only a baby step in delivering a small drop of knowledge. With this note, on behalf of my co-authors, I wish the readers a wonderful reading experience and successful knowledge gain.

(Partha Pratim Sarkar)

Author biographies

Dr Ayan Chatterjee

Dr Ayan Chatterjee received an MTech in Communication Engineering from University of Kalyani in 2012. He obtained a PhD in Microwave Engineering from Indian Institute of Engineering Science and Technology (IIEST), Shibpur in 2018. From 2012 to 2013 he worked as Assistant Professor in the Department of ECE at SKFGI, West Bengal. From 2018 to 2021 he served the Department of ECE at National Institute of Technology, Sikkim. Presently he is working as an Associate Professor in the Department of ECE at University of Engineering and Management, Kolkata. He has authored and co-authored more than 70 papers in referred journals and conference proceedings. His research interests include planar and non-planar frequency selective surfaces, metamaterial inspired surfaces, wideband antennas and antennas for wearable technology. He received a Senior Research Fellowship from CSIR, Government of India in 2014. He has received a Research Grant from DST-SERB in 2023 under the SURE scheme. Dr Chatterjee is actively volunteering at several reputed journals, such as *IEEE Transactions on Antennas and Propagation*, *IEEE Transactions on Microwave Theory and Technology*, *IEEE Access*, *IEEE AWPL*, *Journal of Electromagnetic Waves and Applications*, and many more. He is a Senior Member of URSI as well as IEEE and a Member of Institution of Engineers (India). He is also serving IEEE as the treasurer of the AP/MTT Kolkata Chapter.

Dr Snehasish Saha

Dr Snehasish Saha completed his PhD in engineering and technological studies from University of Kalyani in 2022. His area of specialization in PhD is RF and Microwave Engineering, especially reconfigurable FSS, antenna, and absorber design. He obtained his MTech from University of Kalyani in 2014 in Communication Engineering. He earned his BTech degree in Electronics and Instrumentation Engineering from Department of Engineering and Technological Studies (DETS) in 2012. He has almost 5 years of teaching experience and about 3 years of research experience, excluding the PhD tenure. He is presently working as senior research associate in a NHPC funded project since June 2022 at the Electrical Engineering Department, NIT Durgapur. He has contributed around 17 research articles in various journals and conferences of repute. Among them, 7 research articles are in SCI-indexed journals.

Dr Sushanta Sarkar

Dr Sushanta Sarkar was awarded his PhD by University of Kalyani in 2015. He has studied for a Diploma in Electronics and Telecommunication Engineering from MBC Institute of Engineering and Technology, Burdwan. He obtained his BE in Electronics and Telecommunication Engineering in 2008. In 2010, he completed his MTech from Kalyani Government Engineering College. Presently he is working as an Assistant Professor at the University of Kalyani.

Dr Partha Pratim Sarkar

Dr Partha Pratim Sarkar was awarded a PhD in engineering from Jadavpur University in 2002. He obtained his ME from Jadavpur University in 1994. He earned his BE degree in Electronics and Telecommunication Engineering from Bengal Engineering College (presently known as Bengal Engineering and Science University/ Indian Institute of Engineering Science and Technology, Shibpur) in 1991. He was in the rank of Reader during the period January 1997 to January 2005. He is presently working in the rank of Professor (since January 2005) at the Department of Engineering and Technological Studies, University of Kalyani. His area of research includes microstrip antennas, microstrip filters, frequency selective surfaces and artificial neural networks. He has contributed around 376 research articles in various journals and conferences of repute. Among them, 250 research articles are in peer-reviewed international journals and 132 research articles are in SCI-indexed journals. He has supervised 26 PhD students, 77 MTech thesis and more than 100 BTech projects. He is also a life Fellow of IETE and IE (India).

Chapter 1

Introduction to metamaterial and metasurface

Artificially structured materials that have special electromagnetic characteristics are known as metamaterials. Their structures are designed to demonstrate features that are not naturally attainable. Because of their distinct and tunable effective characteristics, such as magnetic permeability and electric permittivity, metamaterials are essential to the creation of meta-devices. As a result, the majority of current research has been directed at achieving sensing, switchable, nonlinear, and adjustable functionalities. We provide an overview of the most current developments in photonic, microwave, terahertz, and electromagnetic metamaterials as well as their uses in this article. Additionally covered in the review is how metamaterials have advanced microwave sensors, photonic devices, and antennas.

1.1 Introduction

The word 'meta' means 'beyond' in Greek. A novel family of manmade materials known as metamaterials possesses peculiar electromagnetic properties not found in natural materials. Each material in nature has its unique characteristics, be it chemical, mechanical, physical, electrical or based on any other phenomenon. Each material has its own nature. The objects made from the material show characteristics of the materials from which they are made, provided the materials are in the same state or form as the object. But many times it has been observed that, after forming an object from a material, it shows completely different characteristics from the material itself. The phenomenon drew the attention of many researchers. Since then, these objects bearing special unobserved characteristics that didn't occur in its pure material form have become known as metamaterials.

Scientists and engineers began looking for different applications using these types of materials. These metamaterials showed some 'impossible' characteristics.

doi:10.1088/978-0-7503-5422-6ch1

A planar smooth hard surface generally reflects acoustics waves. But imagine, a properly engineered surface made of the planar smooth hard surface that traps the incident acoustics waves within the surface! Such a surface shall act as absorber, even though the material used is a reflector of sound. We can say 'an acoustic-absorbing surface made with acoustic-reflecting materials'—that's a classic example of a metamaterial. In the case of optical metamaterials, imagine an object that bends the light along its surface and meets back at the other side of the object. Generally, the optical light passes through or reflects back to the incident side. But if the surface of the object is engineered so, the object itself becomes invisible as the incident light shall travel to the other side instead of passing through it or reflecting back; it shall take the path around and shall appear again, thus, giving the illusion of an invisibility cloak. Similar concepts can also be applied to electromagnetic metamaterials. Such electromagnetic metamaterials can be used to cover objects that need to be made invisible to detection by radar.

These are few of the many physical attributes that are engineered to get a metamaterial. The engineers can choose any combination of the physical attributes that need to be engineered to obtain the desired type of metamaterial. Complex electromagnetic or microwave devices may need multiple physical properties to be engineered on to metamaterials. The properties of the base materials that are generally engineered to obtain electromagnetic metamaterials are: permittivity, permeability, absorption/reflection of waves, and phase shifting of E field/H field. Natural materials with positive refractive index, magnetic permeability, and electric permittivity include glass, polymer, teflon, polyethylene and many more. In contrast, the new designed materials exhibit negative magnetic susceptibility, negative electric permittivity and permeability leading to a negative value for refractive index. Metamaterials are given various other nomenclatures such as negative index materials (NIM), left-handed (LH) materials, double negative (DNG) media or media with a backward wave (BW).

1.2 Fundamentals of metamaterial and metasurface

Conventional materials, often known as double positive (DPS) media with both epsilon and mu positive ($\varepsilon > 0$, $\mu > 0$), allow right-handed propagation of electromagnetic wave propagation as depicted in first quadrant of figure 1.1. Metamaterials, on the other hand, possess negative values for epsilon and/or mu as shown in the second, third and fourth quadrants in figure 1.1.

Metamaterials, being engineered structures, function in a different manner than materials available in nature. Metamaterials exhibit some unusual properties. These are the structures that reveal zero or negative permittivity with $\varepsilon < 0$ and/or zero, or negative permeability with $\mu < 0$; however, only if effectively engineered [2]. Among many of their unusual properties, based on the above mentioned, there can be three categories of metamaterials.

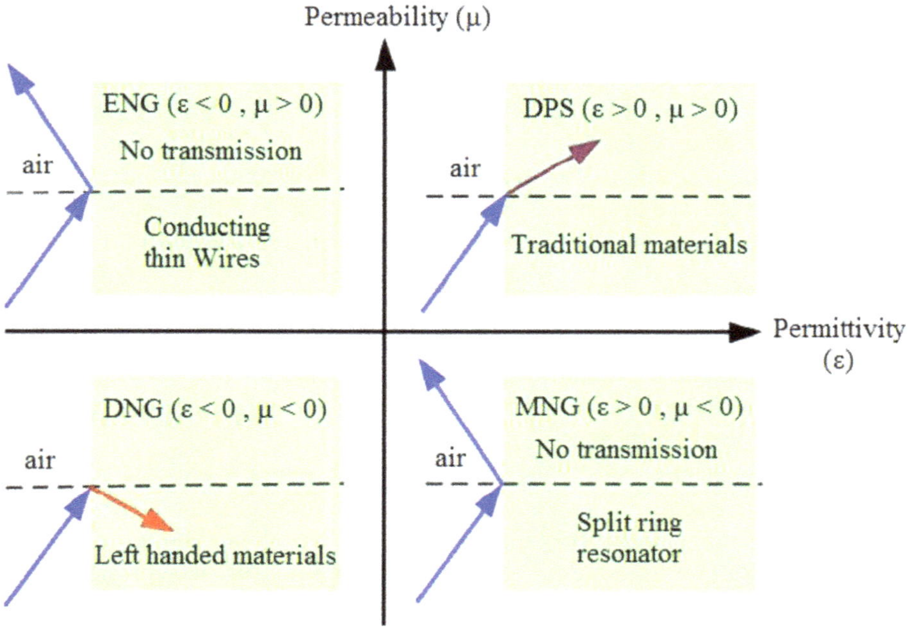

Figure 1.1. Metamaterial inspired structure with sub-wavelength periodicity. Adapted from [1].

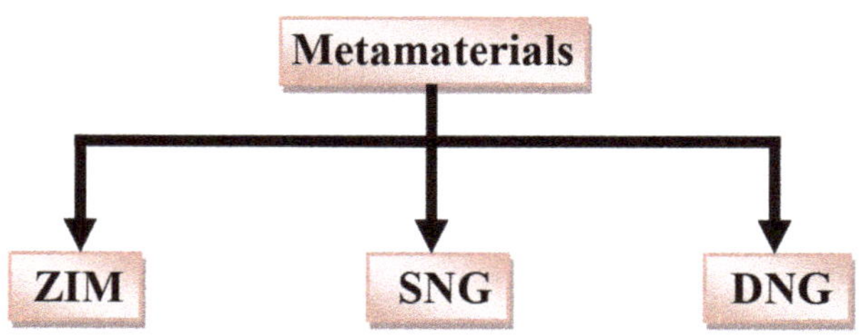

Zero-index materials (ZIMs) possess zero permittivity or zero permeability. Single-negative metamaterials (SNGs) exhibit negative permittivity ($\varepsilon < 0$) or negative permeability ($\mu < 0$). Double-Negative metamaterials (DNGs) possess negative permittivity as well as negative permeability simultaneously.

Victor Veselago carried out the pioneering study of DNG metamaterials theoretically in the year 1967 and also proposed some of their non-conventional properties, such as negative refractive index ($n < 0$), and left-handed propagation of electromagnetic waves [3]. Conventional materials are responsible for the wave propagation in the forward direction, whereas DNG metamaterials are known for backward-wave propagation with the phase velocity and group velocity anti-parallel ($v_p v_g < 0$). The electromagnetic wave travels through such a medium anti-parallel to the pointing vector and consequently the medium is popularly termed left-handed.

Sir John Pendry and his co-researchers investigated the existence of negative permittivity in the microwave spectrum in reality by using a periodic array of thin

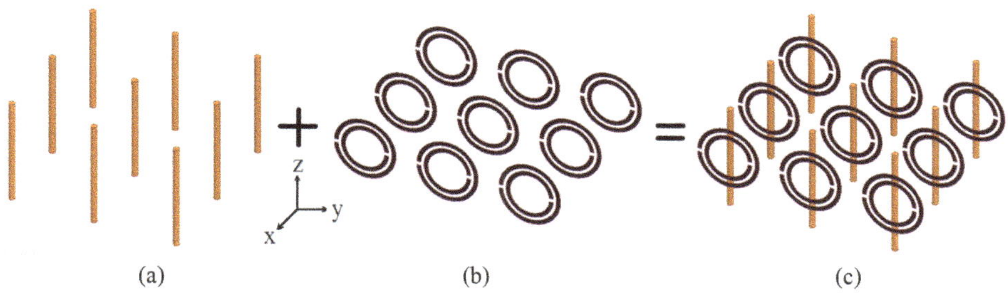

Figure 1.2. Periodic arrangement of (a) thin conductive wires exhibiting $\varepsilon < 0$ (b) split ring resonator exhibiting $\mu < 0$ (c) combined thin metallic wires and SRRs for both $\varepsilon < 0$ and $\mu < 0$.

metallic wires of diameter '*a*' and periodicity '*p*' as shown in figure 1.2(a). This periodic structure is the counterpart of plasmas at the lower end of the microwave spectrum and its effective permittivity falls below zero for an incident electric field parallel to the wires below the plasma frequency [4]. Apart from negative permittivity, they also proposed a structure that can exhibit negative permeability. They came up with an arrangement of metallic patches known as a Split Ring Resonator (SRR), as shown in figure 1.2(b), which exhibits negative permeability at a frequency close to its resonating frequency [5]. He showed that if a time-varying magnetic field is applied perpendicular to the surface of the SRR, then this magnetic field induces current on both the metallic inner and outer rings and consequently electric charge gets build up across the gaps between the rings. The wires and SRRs can be placed together in a three-dimensional periodic arrangement as shown in figure 1.2(c) in order to realize a DNG structure ($\varepsilon < 0$ and $\mu < 0$).

The thin wires can also be designed in the form of metallic strips integrated on a dielectric substrate opposite to the SRRs, as shown in figure 1.3. Besides the non-conventional behaviour it is of great significance that both the metallic wires as well as SRRs are much smaller compared to the wavelength and thus the proposed resonators are sub-wavelength periodic structures. As a result, these structures can be modeled by an equivalent LC circuit whose LC values can also be determined. A single SRR or thin metallic wire shaped unit cell is arranged in large numbers in a periodic manner in order to create an effective epsilon (ε) negative or mu (μ) negative response. With the combination of these two, DNG structures can be realized, as shown in the fabricated prototype of the metamaterial array in figure 1.4 [4].

Metamaterials are structured as periodic arrangements of unit cells, and this is accomplished so as to average the fields across the unit cells and manipulate the structure as a homogenous medium [5]. They are engineered in a way so that the unit cell dimension, also known as periodicity of the unit cell '*p*', is significantly smaller than λ_g, guided wavelength of the microwave ($p \ll \lambda_g$).

The conception of metamaterials was initially theoretically introduced by Russian physicist Victor Veselago in 1968 [3]. He investigated the material's electrodynamics characteristics with negative standards of relative permittivity (ε) and magnetic permeability (μ). Furthermore, included in such materials is the topic of wave propagation at higher frequencies. However, Smith and his research group effectively revealed a structure that demonstrates refraction with negative index in the

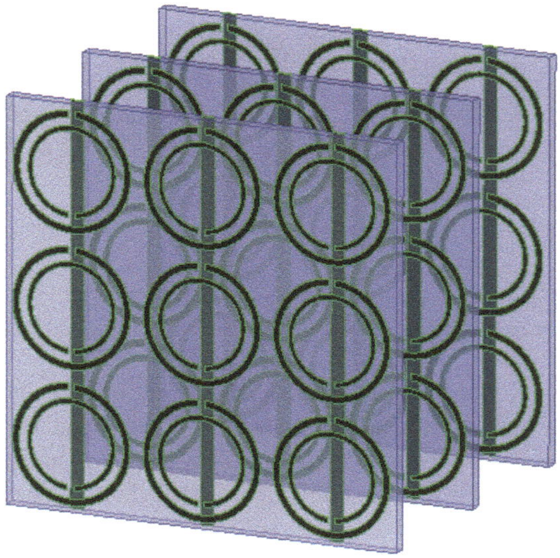

Figure 1.3. Pictorial representation of a DNG metamaterial.

Figure 1.4. Periodic array of thin conductive wires and split ring resonators (SRR). Reprinted from [4], with the permission of AIP Publishing.

microwave regime for the first time [5]. The first ever metamaterial was created by J B Pendry and his research group [4] using two subsystems interpenetrated together. In particular, they created split ring resonators (SRRs) and metallic wires, which have negative permeability and/or permittivity values, by utilizing a variety of thin copper wires and split loaded rings [4, 6, 7]. Metamaterials exhibit a number of unique characteristics, including that of the perfect lense effect [2], reverse Doppler effects [8], the cloaking or invisible effect [9], magnetism at high end of the spectrum [10], vibrant modulation of radiation at terahertz (THz) spectrum [11], and transparent properties induced electromagnetically [12–15]. They are able to create functioning equipment with switching and reconfiguring capabilities that are possible because of the extraordinary qualities of metamaterials [1, 16–19].

Although metamaterials exhibit ENG or MNG responses, on the other hand metamaterial-inspired structures should not necessarily exhibit '$\varepsilon < 0$' or '$\mu < 0$'.

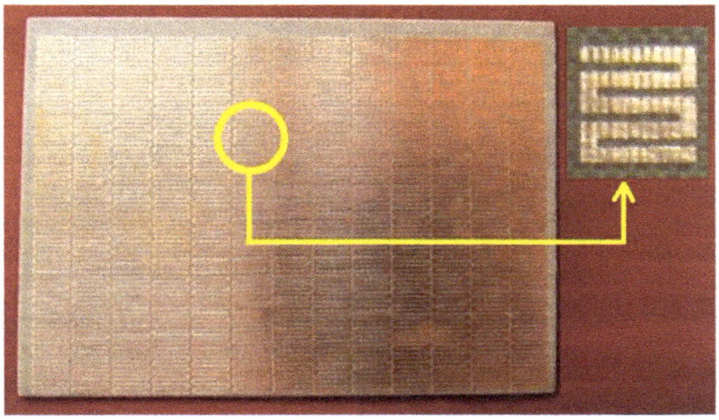

Figure 1.5. Metamaterial inspired structure with sub-wavelength periodicity. Reproduced from [20]. CC BY 4.0.

Most of the metamaterials inspired are two-dimensional periodic structures with a unit cell dimension typically on the order of $\lambda/10$ or less, as shown in figure 1.5 [20]. Periodic structures with sub-wavelength dimension such as electromagnetic bandgap structures (EBGs), frequency selective surfaces (FSSs), and artificial magnetic conductors (AMCs), metasurfaces can be regarded as metamaterial-inspired structures [7], negative (DNG) media or media with backward wave (BW).

In order to exhibit a negative refractive index, three-dimensional metamaterials are required to be realized as shown in figure 1.4. However, such structures possess increased design complexity, high power loss and difficulty of integration with ultrathin devices and applications due to bulky structure. The two-dimensional counterpart of these metamaterials is realized using a different kind of structure known as metasurfaces that can be artificially engineered easily compared to 3D metamaterials with less bulky structure. Metasurfaces can be integrated to planar and low-profile microwave components such as couplers, filters, mixers, and antennas to improve their characteristics, such as bandwidth, efficiency and gain. Moreover, metasurfaces have distinctive capabilities of absorbing, blocking, dispersing, concentrating or guiding electromagnetic waves both in space at normal and oblique incidence as well as on the surface at grazing incidence.

Metasurfaces are a two-dimensional periodic arrangement of artificially engineered unit cells with very small periodicity and are quite similar to frequency selective surfaces (FSSs). This periodic arrangement often exhibits high impedance on its surface. When a plane wave falls on a surface, transmission and reflection both are governed by the impedance of the surface. Metasurfaces are often composed of periodic arrangement of metallic patches with sub-wavelength periodicity backed by a fully conducting ground plane. A resistive metasurafce placed above a dielectric substrate can be seen in figure 1.6 [21].

Metasurfaces designed as high-impedance surfaces can also be used to manipulate surface waves. Surface waves cause serious problems in designing microstrip patch antennas that lead to narrow impedance bandwidth. Metasurfaces thus can be incorporated in patch antennas in order to reduce the surface wave strength, thereby increasing the -10 dB impedance bandwidth of the antenna. Metasurfaces can also be designed by incorporating a conducting ground plane below the array of patches,

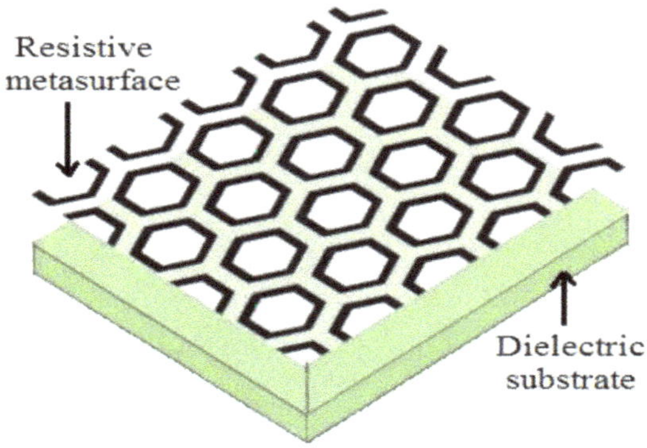

Figure 1.6. A resistive metasurface composed of hexagonal loop shaped patches. Reproduced from [21]. CC BY 4.0.

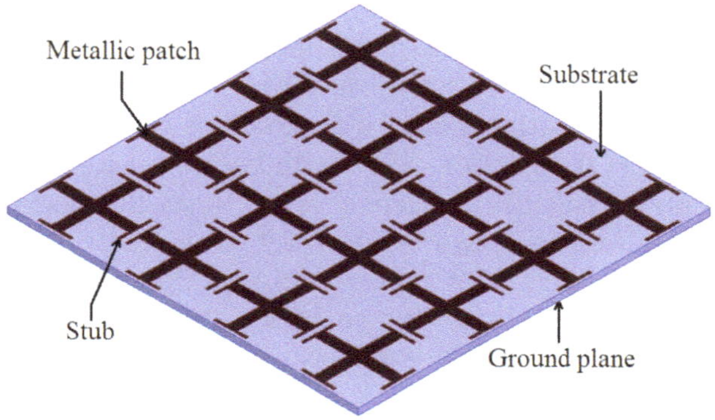

Figure 1.7. A metasurface with Jerusalem cross-shaped patches backed by a ground plane.

as shown in figure 1.7. In such cases the metasurface exhibits absorbing properties. A metasurface with a ground plane shows zero transmission of the waves through it, whereas the reflection also gets reduced, leading to confinement of the waves in the substrate followed by radiation as heat.

1.3 Applications of metamaterial and metasurface

Three-dimensional metamaterials and their two-dimensional counterparts have numerous applications in the field of electromagnetics, solid state physics, classical optics, antenna engineering, optoelectronics, material sciences, and nanotechnology because of their unusual properties as mentioned in the previous section. Some of the significant applications relevant to microwave, mm-wave and beyond are discussed in this section.

1.3.1 Metamaterials in antenna design

The transition from wired to wireless communication systems is occurring in the modern period. In this sense, antennas are required for wireless signal transmission. As a result, there is a greater need for antennas with a large operational bandwidth and a high gain. Researchers are looking for novel types of antennas to meet the needs of contemporary communication. Due to its innovative characteristics, including its wide bandwidth and great radiation efficiency, the dielectric resonator antenna (DRA) garnered a lot of interest in this context [22–24]. DRAs were first the subject of experimental investigations by Long and his colleagues, and other researchers soon followed. Today, DRAs come in a variety of shapes, including rectangle, cylinder, cross, sphere, and hemisphere [25–28]. Rectangular DRAs stand out from the others with their striking qualities. However, the two main disadvantages of DRAs are their limited bandwidth and high manufacturing costs. Cost-effective additive manufacturing, such as 3D printing, can be used in place of expensive conventional manufacturing to lower the fabrication cost. In addition to being economical, 3D printing enables the advancement of intricate structures with particular antenna features for modern and state-of-the art applications.

A star-shaped DRA was recently designed and developed by A Kumar and his co-researchers [29] with the aid of split ring resonator-based metamaterial as superstrate, a programmable logic array (PLA) that is biodegradable, and an affordable 3D printing method as depicted in figure 1.8(a). They revealed that the process of additive manufacturing, which makes use of renewable energy sources, produces PLA, a unique kind of polymer. Furthermore, SDRA exhibits a large bandwidth up to 37%, which might be achieved by reshaping rectangular DRA. Additionally, the circular SRR might act like a metamaterial to increase gain; the antenna can achieve a maximum gain enhancement of 5.4 dB leading to 82.7%. Furthermore, the suggested antenna exhibits circular polarization characteristics at the frequency of 5.8 GHz and shows a radiation efficiency on an average of 80.51% within the operating band, as shown in figure 1.8(b). As a result, the results demonstrate that the suggested antenna satisfies the 5.15–5.35 GHz and 5.725–5.825 GHz WLAN bandwidth requirements. This is helpful for satellite navigation systems, C-band radio navigation for aviation and weather, and communication.

Metamaterials are frequently employed to minimize antenna size so as to achieve frequency response with multiple frequency bands [30]. Wireless vehicle communication systems can greatly benefit from the development of miniaturized antennas,

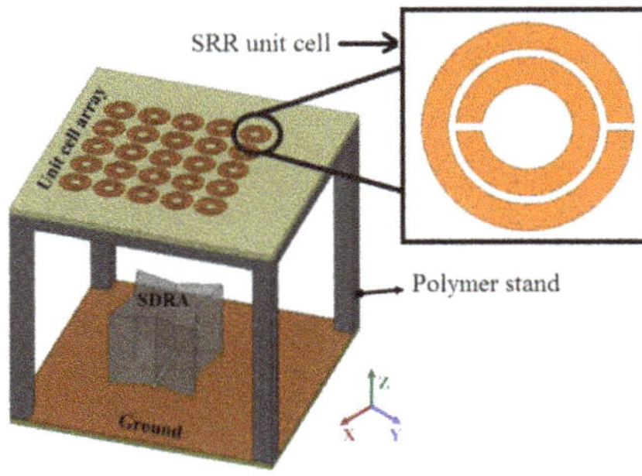

(a)

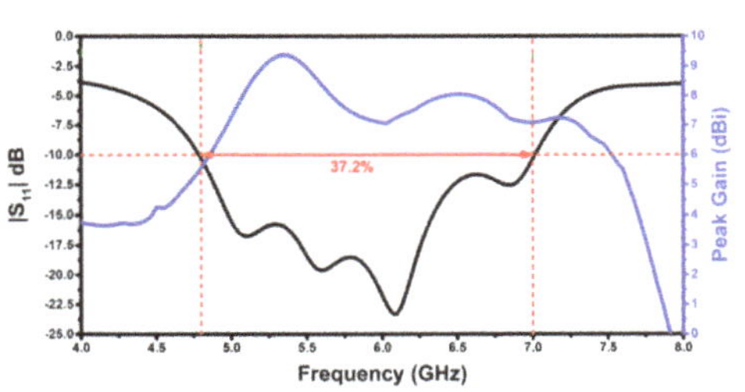

(b)

Figure 1.8. (a) SDRA with the SRR shaped metamaterial-based superstrate. (b) Reflection coefficient and peak gain of the SDRA with the metamaterial-based superstrate [29]. John Wiley & Sons. [© 2019 Wiley Periodicals, Inc.].

that might be acquired through the use of resonances of negative order and zeroth order (ZOR) [31–34]. The development and utilization of monopole antennas loaded by ZOR structure and complementary split ring resonators (CSRRs) were illustrated by Mehdipour and co-researchers [35]. Three tunable frequency bands are available for use with the suggested antennas. They claimed that by packing the zeroth order resonators, the suggested antenna can be made smaller. They also noticed a strong correlation between the outcomes of the simulation and the experiment. Therefore, the proposed antenna's multiband tunable response and compact design demonstrated that it is useful to be implemented in systems used for vehicular communication.

Elwi [36] has presented a new cylindrical antenna designed for multiple input and multiple output systems (MIMOs), featuring a compact structure. The suggested antenna is composed of a cylindrical photo paper (Kodak made) used for a substrate (height $\lambda_o/4.5$ and diameter $\lambda_o/4.5$, where λ_o is the 2.25 GHz free space wavelength) on which four omega-shaped monopoles are folded, with a $\lambda_o/29$) space between each monopole, as shown in figure 1.9. The utmost coupling among the monopole units occurs in the operating band of 2–3 GHz. The SRRs are mounted in the middle of the radiators (monopole) to reduce coupling. The substrate is printed with silver nanoparticles using the ink-jet deposition process. Lastly, it is noted that the suggested array exhibits broad radiation patterns and a gain of 2.5 dB. Numerous technical approaches, such as the use of fractal geometry, shorting walls, shorting pins, and high-permittivity dielectric substrates, have been developed to create compact antennas. These techniques' poor efficiency, low gain, limited bandwidth, and complicated design are their disadvantages. The primary drawback of small or low-profile planar antennas is low gain, which needs to be addressed in order to meet the transmitters' total link budget.

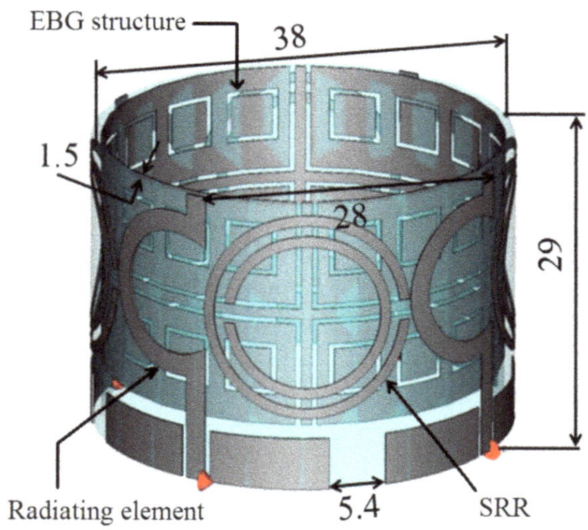

Figure 1.9. Miniaturized folded MIMO antenna array with SRR metamaterial. Reprinted by permission from Springer Nature Customer Service Centre GmbH: Springer Nature, [36], Copyright (2017).

In order to realize high performance antennas for system-on-chip (SoC), metamaterials (MTMs), metasurfaces (MTSs), and substrate integrated waveguide (SIW) techniques have been used more recently. This is possible because of the benefits that these technologies provide, including reduced size, increased bandwidth, and improved radiation properties. Additionally, they prevent surface wave excitation, which could deteriorate the antenna's radiation properties. Additionally, they facilitate the antenna's integration with microwave integrated circuits (MIC). The suggested methods make it easier to integrate and realize SoC antennas on a variety of high-permittivity dielectric substrates, such as polyimide, silicon, graphene, and GaAs, which can lower conducting layer losses and enhance antenna performance.

The ability of metamaterial transmission lines (TLs) to control the characteristics impedance and phase constant set them apart from conventional TLs. As a result, MTMs/MTSs are being used widely in antenna design. Furthermore, we are able to develop innovative antenna structures thanks to the unique properties of MTMs/MTSs, which are not possible with conventional technologies. Studies reveal that MTM/MTS-based antennas have longer operating bandwidths and higher radiated power. When considering antenna designs, the main advantages of MTMs and MTSs over traditional technologies are (i) the ability to realize smaller antennas, (ii) larger bandwidth, (iii) improved radiation characteristics, and (iv) the ability to realize multiband functionality.

On-chip antennas are frequently built on substrates with multiple layers. These antennas are very intricately designed and made. The loss in the substrates and surface wave interactions between the radiating elements make these devices lossy as well. Substrate integrated waveguide technology has proven to be an effective way to minimize substrate and surface wave loss as well as reduce these interactions in on-chip antenna design. Through-the-layer via-pins or via-holes are used in SIW to frame the antenna. This separates individual radiators and lowers loss.

Antenna design at microwave frequencies has made extensive use of MTM, MTS, and SIW technologies. At mm-wave and THz frequencies, they are less common, though these days, research in mm-wave and THz technologies is regarded as one of the most promising and challenging owing to their ability to exhibit elevated data rate required in modern communications and, consequently, high-resolution imaging. On the other hand, creating THz sources is difficult. An approach of designing on-chip THz antennas is suggested by Chun-Hsing Li and Te-Yen Chiu as shown in figure 1.9 [37]. An on-chip antenna is designed using 0.18-μ CMOS technology, and a dielectric resonator antenna is chosen for this purpose as DRA exhibits very small conduction loss at such a high frequency compared to patch antennas. However, this increases the fabrication complexity, which makes the design difficult at THz frequencies. The most recent research on on-chip antennas utilizes MTM and MTS structures for the antenna gain enhancement.

1.3.2 Metamaterial inspired microwave sensor

Liquid characterization and quantification are now essential in many fields, including pharmaceutical, agricultural, and biomedical engineering [38–41].

Generally speaking, the polarity and unique electrical properties of different liquids might be utilized to analyze their properties. Furthermore, the electrical properties of the liquids have a significant effect on the microwave devices' performance. Additionally, the polarization direction of various molecules varied as a result of the electromagnetic radiation's coupling with the polar liquid materials. This kind of interaction is specifically used by microwave sensors to modify the liquids' dielectric characteristics for characterization. The electromagnetic approach-based sensors are superior to conventional ones in a number of ways because of their straightforward operation, non-invasive nature, and speedy response.

Furthermore, during the past 20 years, the development of electromagnetic devices has focused a lot of attention on electromagnetic metamaterials due to their special qualities. Electromagnetic metamaterial-based structures are currently being implemented in microfluidic sensing applications [42, 43]. These sensors show a well-built coupling of the analytes with the electric field in extension to their high sensitivity. However, a significant drawback of microfluidic sensors is their requirement of a large sample volume. Implementation of structures resonating in nature in the proposed microfluidic sensors can lessen this [44, 45]. However, obtaining the necessary microlevel sensitivity remains a significant challenge. In order to get around this, a lot of researchers are working non-stop to create tiny sensors with high selectivity and sensitivity for reliable liquid characterization. Continuing on from here, Xu *et al* [46] presented a biocompatible, lightweight, low-cost, portable, flexible, and metamaterial-inspired photonic structure for application in strain with high sensitivity along with biological as well as chemical sensing. Both infrared and invisible modes are available for the device. The suggested device consists of U-shaped split ring resonators made of silver (Ag) or gold (Au) with a thickness of 30 nm, that are set down using electron beam lithography on a poly (ethylene-naphthalate) substrate [47]. Furthermore, the proposed SRR based metamaterials exhibit a 542 nm electric resonance and a 756 nm magnetic resonance. Both electric and magnetic resonant modes respond extremely sensitively to surface chemical environments, bending strain, and surrounding dielectric media. Furthermore, it exhibits an enhanced reaction for a self-assembled monolayer of 2-naphthalenethiol, exhibiting a 65 nm shift in magnetic resonance. These results imply that the suggested instrument is a strong contender for biological and chemical sensing. Electromagnetic devices made of synthetic materials are called microwave metamaterials. Once these materials are interfaced with the human body, they can be utilized in future healthcare systems that are able to overcome technological limitations.

The principles and applications of microwave metamaterials for biomedical sensing were investigated by John S Ho *et al* [48]. Additionally, Kayal *et al* [49] used mu negative (MNG) metamaterials to demonstrate a compact microwave sensor for liquid characterization. They stated that the prepared sensor is observably compact and has a high sensitivity. In order to achieve both notable compactness and sensitivity in small cross-sectional areas, the square spiral metamaterial (MNG) is crucial. Furthermore, the prepared device's sensing behavior is verified using the least squares technique, and two nonlinear equations are then developed for calibration purposes. Equations (9) and (10) of reference [49] are nonlinear equations that can be used to determine the permittivity of unknown samples. Therefore, the prepared sensor is a strong contender for liquid sensing applications due to its compact design and high sensitivity.

The electromagnetic radiation in the terahertz (THz) regime exhibits reduced photon energy (in the range of meV) and susceptible reaction to inter- and intra-molecular trembling modes [50, 51]. As a result, electromagnetic radiation at THz frequency shows promise for use in spectroscopy, microscopy, and biosensing. Using graphene and tens of metamaterials, Xu and the research group [52] presented a metamaterial that can sense biomolecules in the THz region. Additionally, Lee *et al* [53] established a label-free sensing method for the discrimination of mono-stranded deoxyribonucleic acids (ssDNAs) at the THz region using graphene-assisted nano metamaterials as shown in figure 1.10.

The exclusive properties of metamaterials combined with the electro-optical qualities of graphene allow for biomolecule sensing, even THz photons of very small energy levels. Furthermore, they demonstrated that the graphene sheet's cross-sectional absorption rises in response to the augmentation of the terahertz radiation at resonating frequency, offering ultrahigh sensitivity. The embattled DNA molecules are tightly bound without undergoing structural modifications by the implementation of graphene onto a metamaterial made of nano-slots, which is a significant candidate of the sensing mechanism. In this case, the THz transmission capability is increased by the proposed metamaterials in proportion to the cross-section of absorption of the DNA-adsorbed graphene layer as shown in figure 1.11. Therefore, a larger portion of DNA molecules might be seen using electromagnetic radiation at the THz regime that are intensely focused.

Moreover, absorbed molecules alter graphene's easily detectable intrinsic electrical properties. It is possible for this mechanism to detect various biomolecules. In particular, choosing an appropriate receptor to bind DNA molecules, doing a quick primary screening, and then applying for DNA sequencing. Therefore, the sensing domain of metamaterial at THz with graphene assistance is appropriate for

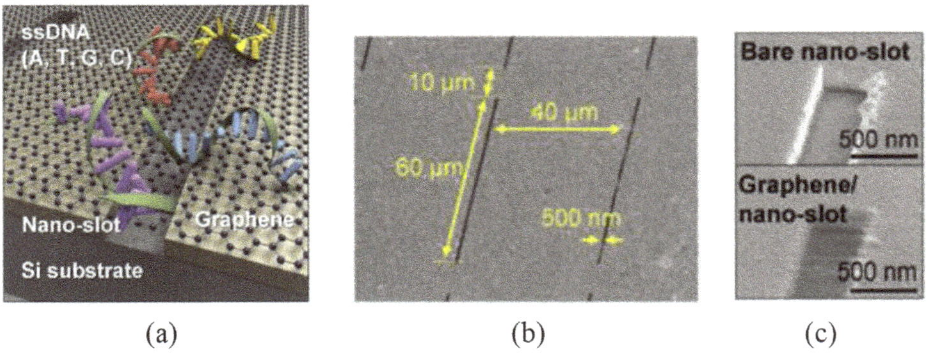

Figure 1.10. (a) DNA adsorption on the graphene-combined nano-slot metamaterial. (b) Geometry of the sensing chip. (c) Scanning electron microscopy (SEM) images for bare and graphene-combined nano-slots. Reproduced from [00]. CC BY 4.0.

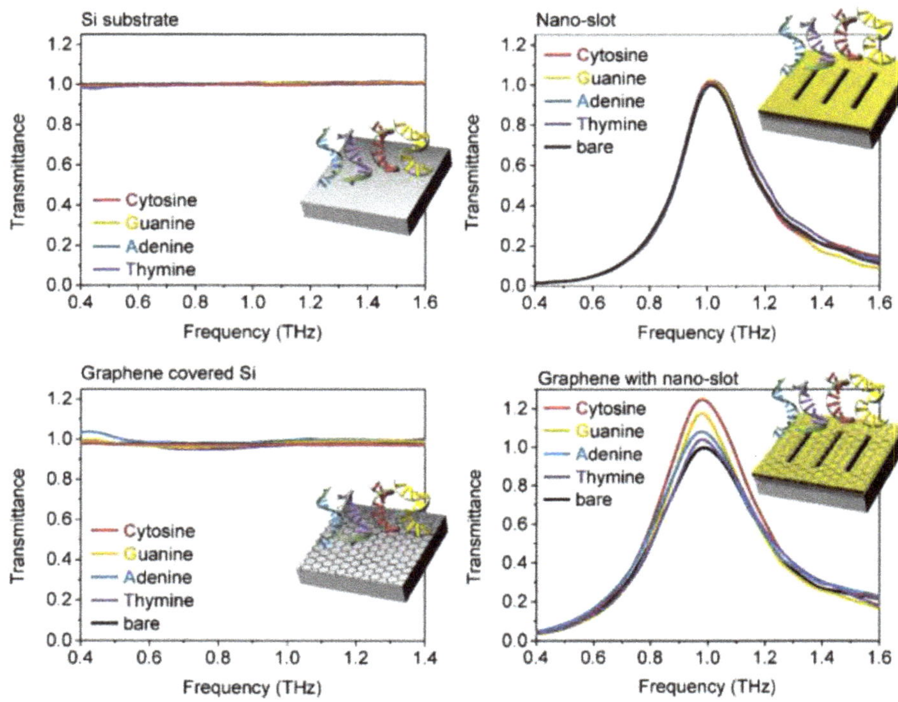

Figure 1.11. Variation in THz transmission response due to ssDNA on silicon, a bare nano-slot metamaterial, graphene-covered Si, and a graphene-covered nano-slot metamaterial. Reproduced from [53]. CC BY 4.0.

biological sensing applications, at the same time for comprehending the electro-optic behavior of two-dimensional materials.

The linking of transmission lines to metamaterial-based open loop resonators can increase the functioning ability of the sensor in the microwave regime [54]. However, a shift in the frequency of resonance occurs when the resonator is coupled to the transmission line, and this has a significant impact on achieving high sensitivity. Abdolrazzaghi and co-researchers [55] suggested a novel microwave sensor of planar configuration based on metamaterials that can function at the frequency of 2.5 GHz so as to get around this. They coupled metamaterials with negative refractive index and transmission lines to create the above-proposed sensor, which has significantly better resonant properties [56]. They then created a signal flow analysis to calculate the prepared sensor's transmission response. The suggested sensor exhibits a complex permittivity of high value including very high sensitivity when compared to the traditional sensors, such as microstrip line-based sensors. They stated that the suggested sensor exhibits better qualities, especially in materials with high permittivity and in the water host medium. The exceptional functionality of the suggested sensor compared to the traditional sensor is revealed by the measurements of concentration of ethanol or methanol contained in the water medium. Therefore, the suggested metamaterial-based microwave sensor of planar configuration is helpful for bimolecular detection, highly sensitive ethanol or methanol in water concentration measurement, and material characterization with high permittivity.

1.3.3 Design of FSS using metamaterial and metasurface

Frequency selective surfaces are a two-dimensional array of metallic patches on a dielectric substrate or that of apertures within a metallic screen. The ground plane on the other side of the substrate is removed. It acts as a band stop or bandpass filter in the microwave spectrum as well as the mm-wave regime. Sometimes it may be designed with the incorporation of a fully conducting ground plane so that it may be used as an absorber or artificial magnetic conductor. Metamaterial-based structures can be incorporated in designing FSS unit cells in order to avail the unusual properties of the metamaterial in FSS characteristics. Moreover, the primary advantage becomes miniaturization of the unit cell of the FSS with the use of non-resonant structures of metamaterials and metasurfaces. In figures 1.12 and 1.13 the structure of a band stop FSS using normal material and its response are shown.

In figures 1.14 and 1.15 the structure of bandpass FSS using normal material and its response are shown. Bandwidth and operating band depend upon the shape and dimension of the unit cell and also upon the periodicities of elements. Numerous research works are going on regarding how the size of the frequency selective surface can be reduced while keeping all the characteristics like bandwidth and operating band as it was before. Design of metamaterial- or metasurface-inspired frequency selective surfaces may be the solution in this regard. Metamaterial-based FSS unit cells exhibit miniaturization on a much smaller order compared to the wavelength at resonance and may extend even up to one tenth of the wavelength of $\lambda/10$. Moreover the structures acquire ultrathin substrates, leading to an overall reduction in structure dimension. In certain cases, instead of using popular metamaterial structures such as SRR, CSRR, etc., metamaterial-inspired unit cells can also be used [58]. In such cases sub-wavelength dimensions are maintained, however with or without epsilon negative and mu negative.

Figure 1.12. Structure of a frequency selective surface with band stop response. [57] John Wiley & Sons. [Copyright © 2007 Wiley Periodicals, Inc.]

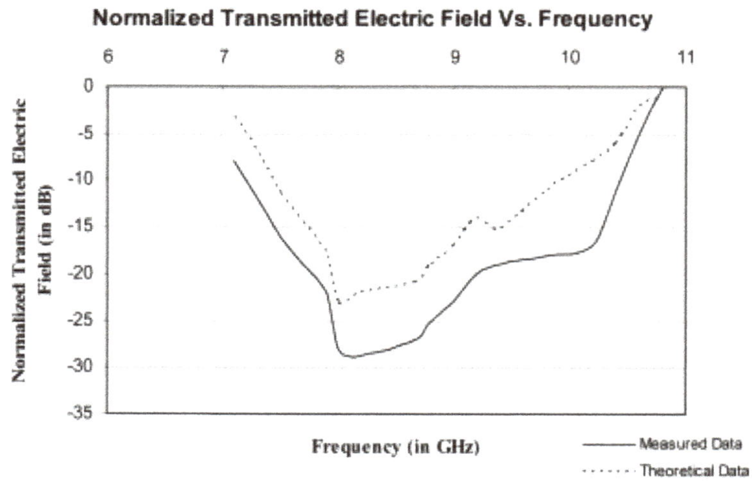

Figure 1.13. Transmission response of the frequency selective surface showing stop band. [57] John Wiley & Sons. [Copyright © 2007 Wiley Periodicals, Inc.]

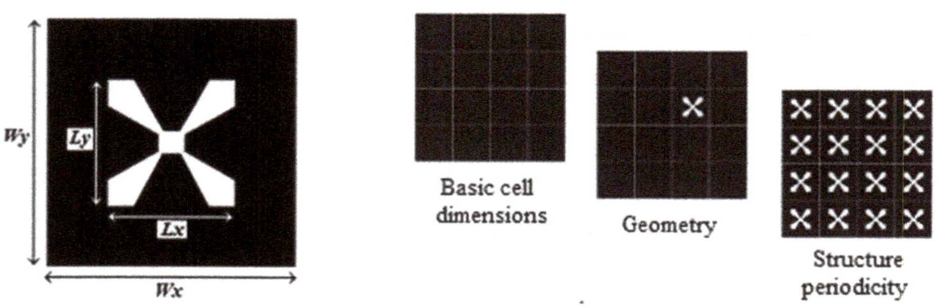

Figure 1.14. Structure of a frequency selective surface with bandpass response. [59] John Wiley & Sons. [© 2016 Wiley Periodicals, Inc.]

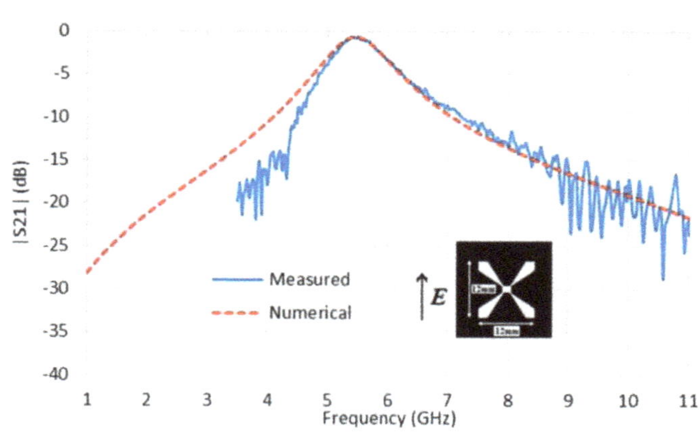

Figure 1.15. Transmission response of the frequency selective surface showing passband. [59] John Wiley & Sons. [© 2016 Wiley Periodicals, Inc.]

1.4 Conclusion

Metamaterials are highly engineered materials that provide properties that are not found in nature. Owing to their distinct characteristics, metamaterials have garnered significant interest across multiple domains such as antennas, photonics, sensors, frequency selective surfaces, energy harvesting, and so forth. Metamaterials based on graphene effectively control electromagnetic wave absorption, transmission, and reflection [20, 21, 37]. Applications involving biological sensing are appropriate for the graphene-assisted THz metamaterial sensing platform. There are still a lot of obstacles to overcome despite the extensive research being done to find new, tunable metamaterials with better performance.

References

[1] Engheta N and Ziolkowski R 2006 *Metamaterials: Physics and Engineering Explorations* (New York: Wiley)

[2] Pendry J B 2000 Negative refraction makes a perfect lens *Phys. Rev. Lett.* **85** 3966

[3] Veselago V G 1967 The electrodynamics of substances with simultaneously negative values of ε and μ *USP. Fiz. Nauk* **92** 517–26

[4] Wilson J D and Schwartz Z D 2005 Multifocal flat lens with left-handed metamaterial *Appl. Phys. Lett.* **86** 021113

[5] Smith D R, Padilla W J, Vier D C, Nemat-Nasser S C and Schultz S 2000 Composite medium with simultaneously negative permeability and permittivity *Phys. Rev. Lett.* **84** 4184–7

[6] Lalas A X, Kantartzis N V and Tsiboukis T D 2016 Metamaterial-based wireless power transfer through interdigitated SRRs *COMPEL: Int. J. Comput. Math. Electr. Electron. Eng.* **35** 1338–45

[7] Pendry J B, Holden A J, Stewart W J and Youngs I 1996 Extremely low frequency plasmons in metallic mesostructures *Phys. Rev. Lett.* **76** 4773–6

[8] Shelby R A, Smith D R and Schultz S 2001 Experimental verification of a negative index of refraction *Science* **292** 77–9

[9] Schurig D, Mock J J, Justice B J, Cummer S A, Pendry J B, Starr A F and Smith D R 2006 Metamaterial electromagnetic cloak at microwave frequencies *Science* **314** 977–80

[10] Linden S, Enkrich C, Dolling G, Klein M W, Zhou J, Koschny T, Soukoulis C M, Burger S, Schmidt F and Wegener M 2006 Photonic metamaterials: magnetism at optical frequencies *IEEE J. Sel. Top. Quantum Electron.* **12** 1097–105

[11] Li C, Wu J, Jiang S, Su R, Zhang C, Jiang C, Zhou G, Jin B, Kang L, Xu W *et al* 2017 Electrical dynamic modulation of THz radiation based on superconducting metamaterials *Appl. Phys. Lett.* **111** 092601

[12] Papasimakis N, Fedotov V A, Zheludev N I and Prosvirnin S L 2008 Metamaterial analog of electromagnetically induced transparency *Phys. Rev. Lett.* **101** 253903

[13] Kurter C, Tassin P, Zhang L, Koschny T, Zhuravel A P, Ustinov A V, Anlage S M and Soukoulis C M 2011 Classical analogue of electromagnetically induced transparency with a metal-superconductor hybrid metamaterial *Phys. Rev. Lett.* **107** 043901

[14] Jin B B, Wu J B, Zhang C H, Jia X Q, Jia T, Kang L, Chen J and Wu P H 2013 Enhanced slow light in superconducting electromagnetically induced transparency metamaterials *Supercond. Sci. Technol.* **26** 074004

[15] Zhang C, Wu J, Jin B, Jia X, Kang L, Xu W, Wang H, Chen J, Tonouchi M and Wu P 2017 Tunable electromagnetically induced transparency from a superconducting terahertz metamaterial *Appl. Phys. Lett.* **110** 241105
[16] Zheludev N I 2010 The road ahead for metamaterials *Science* **328** 582–3
[17] Zheludev N I 2011 A roadmap for metamaterials *Opt. Photonics News* **22** 31–5
[18] T X C 2018 *Science* **vol 262** (New York: Springer International Publishing AG)
[19] Capolino F 2017 *Theory and Phenomena of Metamaterials: Metamaterials Handbook* (Boca Raton, FL: CRC Press)
[20] Chatterjee A, Banerjee S, Frnda J and Dvorsky M 2022 Planar FSS based dual-band wire monopole antenna for multi-directional radiation with diverse beamwidths *IEEE Access* **10** 30427–35
[21] Meng F G, Li H, Fan D G, Li F F, Xue F Z, Chen P and Wu R X 2018 Transmitting-absorbing material based on resistive metasurface *AIP Adv.* **8** 075008
[22] Petosa A and Norwood M A 2007 *Artech House Antennas and Propagation Library* (Norwood, MA: Artech House Publishers)
[23] Kumar J and Gupta N 2014 Performance analysis of dielectric resonator antennas *Wirel. Pers. Commun.* **75** 1029–49
[24] Luk K M and Leung K W 2002 *Dielectric Resonator Antennas* (Hertforodshire: Research Studies Press Limited)
[25] Kumar J and Gupta N 2015 Bandwidth and gain enhancement technique for Gammadion cross dielectric resonator antenna *Wirel. Pers. Commun.* **85** 2309–17
[26] Kumar J and Gupta N 2015 Linearly polarized asymmetric dielectric resonator antenna for 5.2-GHz WLAN applications *J. Electromagn. Waves Appl.* **29** 1228–37
[27] Petosa A and Ittipiboon A 2010 Dielectric resonator antennas: a historical review and the current state of the art *IEEE Antennas Propag. Mag.* **52** 91–116
[28] Kumar P, Dwari S and Kumar J 2017 Design of biodegradable quadruple-shaped DRA for WLAN/Wi-Max applications *J. Microw. Optoelectron. Electromagn. Appl.* **16** 867–80
[29] Kumar A, Kapoor P, Kumar P, Kumar J and Kumar A 2019 Metamaterial loaded aperture coupled biodegradable star-shaped dielectric resonator antenna for WLAN and broadband applications *Microw. Opt. Technol. Lett.* **62** 264–77
[30] Dong Y and Itoh T 2012 Metamaterial-based antennas *Proc. IEEE* **100** 2271–85
[31] Gong J Q, Jiang J B and Liang C H 2012 Low-profile folded-monopole antenna with unbalanced composite right-/left-handed transmission line *Electron. Lett.* **48** 813–5
[32] Iizuka H and Hall P S 2007 Left-handed dipole antennas and their implementations *IEEE Trans. Antennas Propag.* **55** 1246–53
[33] Zhu J and George V E 2008 A compact transmission-line metamaterial antenna with extended bandwidth *IEEE Antennas Wirel. Propag. Lett.* **8** 295–8
[34] Herraiz M, Francisco J, Peter S H, Qing L and Daniel S-V 2011 Left-handed wire antennas over ground plane with wideband tuning *IEEE Trans. Antennas Propag.* **59** 1460–71
[35] Mehdipour A, Tayeb A D and Abdel-R S 2013 Multi-band miniaturized antenna loaded by ZOR and CSRR metamaterial structures with monopolar radiation pattern *IEEE Trans. Antennas Propag.* **62** 555–62
[36] Elwi T A 2018 A miniaturized folded antenna array for MIMO applications *Wirel. Pers. Commun.* **98** 1871–83
[37] Li C -H and Chiu T -Y 2017 340-GHz low-cost and high-gain on-chip higher order mode dielectric resonator antenna for THz applications *IEEE Trans. Terahertz Sci. Technol.* **7** 284–94

[38] Banerjee P, Ghosh G and Biswas S K 2010 Measurement of dielectric properties of medium loss samples at X-band frequencies *J. Metall. Mater. Sci.* **52** 247–55

[39] Grenier K, Dubuc D, Poleni P, Kumemura M, Toshiyoshi H, Fujii T and Fujita H 2009 Integrated broadband microwave and microfluidic sensor dedicated to bioengineering *IEEE. Trans, Microw. Theory Tech.* **57** 3246

[40] Lee H J, Lee J H, Choi S, Jang I S, Choi J S and Jung H I 2013 Asymmetric split-ring resonator-based biosensor for detection of label-free stress biomarkers *Appl. Phys. Lett.* **103** 0537021

[41] Rawat V, Dhobale S and Kale S N 2014 Ultra-fast selective sensing of ethanol and petrol using microwave-range metamaterial complementary split-ring resonators *J. Appl. Phys.* **116** 1641061

[42] Gordon J A, Holloway C L, Booth J, Kim S, Wang Y, BakerJarvis J and Novotny D R 2011 Fluid interactions with metafilms/metasurfaces for tuning, sensing, and microwave-assisted chemical processes *Phys. Rev.* B **83** 205130

[43] Awang R A, Tovar-Lopez F J, Baum T, Sriram S and Rowe W S 2017 Meta-atom microfluidic sensor for measurement of dielectric properties of liquids *J. Appl. Phys.* **121** 094506

[44] Withayachumnankul W, Jaruwongrungsee K, Tuantranont A, Fumeaux C and Abbott D 2013 Metamaterial-based microfluidicsensor for dielectric characterization *Sensors Actuators* A **189** 2331

[45] Velez P, Grenier K, Mata-Contreras J, Dubuc D and Martín F 2018 Highly-sensitive microwave sensors based on open complementary split ring resonators (OCSRRs) for dielectric characterization and solute concentration measurement in liquids *IEEE Access* **6** 48324–38

[46] Xu X, Peng B, Li D, Zhang J, Wong L M, Zhang Q, Wang S and Xiong Q 2011 Flexible visible–infrared metamaterials and their applications in highly sensitive chemical and biological sensing *Nano Lett.* **11** 3232–8

[47] Ahn S H and Guo L J 2009 Large-area roll-to-roll and roll-to-plate nanoimprint lithography: a step toward high-throughput application of continuous nanoimprinting *ACS Nano* **3** 2304–10

[48] Ho J S and Li Z 2021 Microwave metamaterials for biomedical sensing *Reference Module in Biomedical Sciences* (Amsterdam: Elsevier)

[49] Kayal S, Shaw T and Mitra D 2019 Design of metamaterial-based compact and highly sensitive microwave liquid sensor *Appl. Phys.* A **126** 1–9

[50] Choi G, Bahk Y-M, Kang T, Lee Y, Son B H, Ahn Y H, Seo M and Kim D-S 2017 Terahertz nanoprobing of semiconductor surface dynamics *Nano Lett.* **17** 6397–401

[51] Lee D-K, Kang J-H, Kwon J, Lee J-S, Lee S, Woo D H, Kim J H, Song C-S, Park Q-H and Seo M 2017 Nano metamaterialsfor ultrasensitive terahertz biosensing *Sci. Rep.* **7** 8146

[52] Xu W, Xie L, Zhu J, Tang L, Singh R, Wang C, Ma Y, Chen H-T and Ying Y 2019 Terahertz biosensing with a graphene metamaterial heterostructure platform *Carbon* **141** 247–52

[53] Lee S-H, Choe J-H, Kim C, Bae S, Kim J-S, Park Q-H and Seo M 2020 Graphene assisted terahertz metamaterials for sensitive bio-sensing *Sensors Actuators* B **310** 127841

[54] Eleftheriades G V, Iyer A K and Kremer P C 2002 Planar negative refractive index media using periodically LC loaded transmission lines *IEEE Trans. Microw. Theory Techn.* **50** 2702–12

[55] Abdolrazzaghi M, Daneshmand M and Iyer A K 2018 Strongly enhanced sensitivity in planar microwave sensors based on metamaterial coupling *IEEE Trans. Microw. Theory Tech.* **66** 1843–55
[56] Ran L, Huangfu J, Chen H, Li Y, Zhang X, Chen K and Kong J A 2004 Microwave solid-state left-handed material with a broad bandwidth and an ultralow loss *Phys. Rev.* B **70** 07302
[57] Pain M K, Bhunia S, Biswas S, Sarkar D and Sarkar P P 2007 A novel investigation on size reduction of a frequency selective surface *Microw. Opt. Technol. Lett.* **49** 2820–1
[58] Tong X C and Xingcun C T 2018 Metamaterials inspired frequency selective surfaces *Functional Metamaterials and Metadevices* (Berlin: Springer) 155–71
[59] Gomes Neto A, Costa e Silva J, Nogueira de Carvalho J, da Costa A P and de Moura L C M 2016 Bandpass frequency selective surface using asymmetrical slot four arms star geometry *Microw. Opt. Technol. Lett.* **58** 1105–9

Chapter 2

Basics of antenna and frequency selective surface

Wireless communication has become one of the most rapidly growing technologies all over the world in the past two decades. It has a wide range of applications ranging from Wi-Fi, WLAN, and television broadcasting to satellites, radar, spacecraft, and navigation systems, most of which are in the microwave regime [1, 2]. One of the significant elements in a wireless communication sub-system is the antenna. However most of the applications have a requirement of high speed and secured communication with adequate signal strength in order to suppress noise and interference [3, 4]. Antennas alone are not capable of meeting these requirements and additional structures are required to be integrated with the antenna. Microwave periodic structures such as electromagnetic bandgap structures (EBGs), frequency selective surfaces (FSSs), artificial magnetic conductors (AMCs), metamaterials, etc., are potential candidates for the planar as well as non-planar antennas [5–7]. This chapter focuses on fundamentals of antennas and FSSs, basic parameters, design challenges and various classes of these structures in the microwave regime along with their applications.

2.1 Fundamentals of antennas

Antennas are used for transmission and reception of signals in a specific direction and with a specific strength. According to the IEEE Standard for Definitions of Terms for Antennas, antenna is 'that part of a transmitting or receiving system that is designed to radiate or to receive electromagnetic waves' [8]. It performs the transition of energy carrying electromagnetic waves between free space and a guiding device. The guiding device may be a transmission line such as coaxial line, microstrip line or a waveguide. All antennas function on the same principles of electromagnetism as proposed by Sir J C Maxwell. However, the journey started while Heinrich Hertz [9], tried to validate Maxwell's theory with his experiments.

doi:10.1088/978-0-7503-5422-6ch2

The first ever radio system operating at 75 MHz consisted of a half-wavelength dipole (transmitting antenna) and a resonant loop (receiving antenna) [10]. The induction coil was turned on and sparks were induced in the gap that was detected at the receiving antenna.

However the first use of antennas for long distance communication was initiated in 1901 when Marconi sent radio waves from a station in Cornwall, England across the Atlantic to St. Johns, Newfoundland [12]. A fan antenna with 50 vertical wires supported by two poles was used in the transmitter, whereas the receiving antenna consisted of a 200 m long wire that was pulled up with a kite. Following this experiment various antennas were developed at the lower end of the RF spectrum and most of them were wire antennas. During World War II depending on the need of the communication systems focus was given on designing microwave antennas with directive beams. Horn antenna has been one of the frequently used microwave antennas since the 19th century.

2.1.1 Radiation from antenna

A waveguide fed horn antenna is shown in figure 2.1, where the waveguide is excited by a microwave source with sinusoidal output [10]. The solid lines (straight) represent electric fields (E) inside the waveguide, whereas the curved lines represent the fields in the antenna aperture and those radiated from the antenna. Three upward field lines (solid) reach maximum signal strength at time $t = T/4$ (T being time period) and travel a radial distance of $\lambda/4$ (λ being the wavelength), whereas in the next quarter they travel another $\lambda/4$ distance and charge density on the antenna begins to diminish. This may be accomplished by introducing opposite charges, which at $t = T/2$ have neutralized the charges on the antenna.

The field lines created by the opposite charges (dashed line) travel a distance of $\lambda/4$ during the second quarter of the first half. At the end there are three upward field lines in the first $\lambda/4$ distance and three downward field lines in the second $\lambda/4$. To assure there is no charge left on the antenna, the field lines must have been forced to isolate them from the antenna and unite together forming closed loops. The process continues indefinitely and the electric fields are radiated from the antenna. Similarly magnetic fields (oriented in the direction orthogonal to electric field) also travel in the form of closed loops leading to radiation of waves from the antenna.

2.1.2 Basic parameters of antenna

An antenna can be characterized by certain fundamental parameters and they can be analyzed primarily from far field radiation pattern of an antenna. These parameters include impedance bandwidth, gain, directivity, efficiency and polarization. Any antenna in the RF spectrum can be distinguished with the help of these fundamental parameters.

2.1.2.1 Radiation pattern

Radiation pattern is a plot of spatial variation of field intensity radiated from an antenna. Generally radiation patterns can be represented in a rectangular plot or

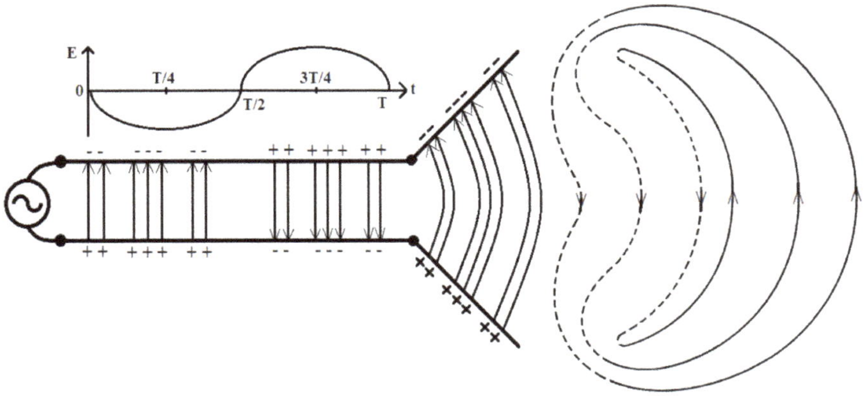

Figure 2.1. Radiation mechanism for a standard antenna showing propagation of electric fields.

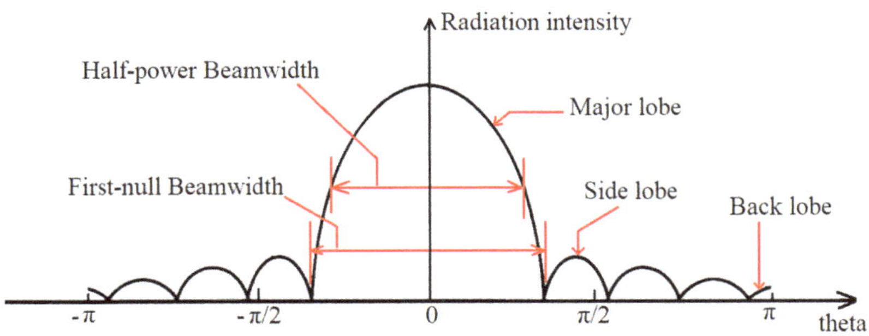

Figure 2.2. Radiation plot of an antenna and various lobes.

polar plot. A rectangular plot of a typical radiation pattern is shown in figure 2.2. The spatial variation corresponds to the variation of the angle theta (θ) for a constant phi (φ) or vice versa [10]. The 3D polar plot of the radiation pattern of an antenna with the angles is shown in figure 2.3(a). Since the radiated signal strength decreases with distance R from the source, radiation pattern should be measured at a specific distance from the antenna. Radiation pattern can be represented by plotting the spatial variation of electric or magnetic field, or it can be a plot of square of the field either in linear scale or dB scale [10]. The radiation plot in figure 2.2 represents variation in radiation intensity, the amount of power radiated from an antenna per unit solid angle.

As shown, a radiation pattern has various levels of radiation intensity over angle (theta) and they are distinguished as lobes. The major lobe has maximum radiation intensity and is generally found in the broadside direction, whereas any lobe other than the major lobe is known as a minor lobe. Generally the lobes adjacent to the major lobe are known as side lobes. A back lobe makes an angle of 180° with respect to the main beam of an antenna. A difference of nearly 20 dB is desirable between the major lobe and side lobe levels for most of the applications [13]. Different beamwidths can be evaluated from the radiation plot. In the main lobe, the angle between the two

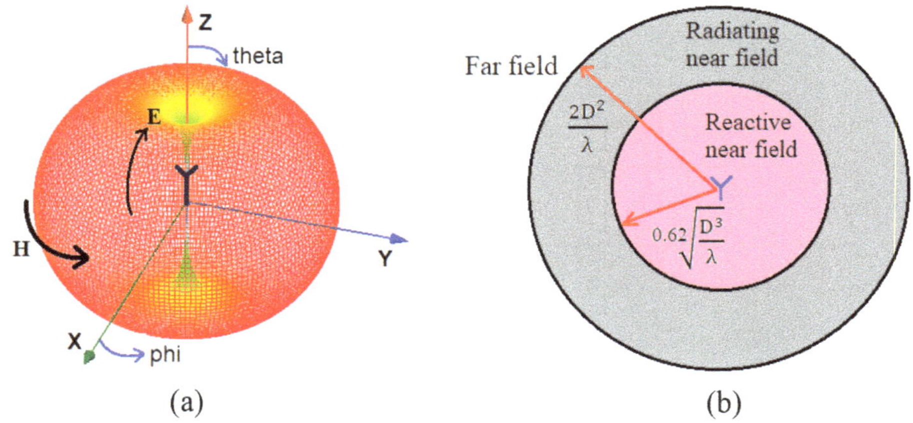

Figure 2.3. (a) Radiation plot of an antenna and various lobes (b) *E* plane and *H* plane in the radiation plot.

directions along which the radiation intensity is half of the maximum value, is termed as half-power beamwidth (HPBW). The angular difference between the first nulls of the pattern is referred to as the first-null beamwidth (FNBW) [14].

The radiation plot depends on the distance *R* between observation point and the antenna. The power pattern is plotted for a constant radius (*R*) and the space surrounding an antenna is subdivided into three regions as depicted in figure 2.3(b).

The region close to an antenna with the radius of $R < 0.62\sqrt{\frac{D^3}{\lambda}}$, where reactive fields predominate is known as reactive near field, in which λ is the wavelength at operating frequency, and *D* is the largest dimension of the antenna [10]. The region with $0.62\sqrt{\frac{D^3}{\lambda}} < R < \frac{2D^2}{\lambda}$ has a predominance of radiating fields and exists only for largest dimension of the antenna $D \gg \lambda$, is known as radiating near field. The region beyond this ($R > \frac{2D^2}{\lambda}$) where angular field distribution is independent of the distance from the antenna, is known as far-field [10].

Due to variation in the magnitude and phase of the fields, the shape of the radiation pattern of an antenna changes as it moves from the reactive near field to the far field. A well-formed radiation pattern can be achieved in the far-field consisting of certain minor lobes and one, or more, major lobes. Moreover, the radiation pattern of an antenna is plotted in two principal planes, namely the *E*-plane and *H*-plane [14]. The plane composed of the electric field vector and the direction of maximum amount of radiation is known as the *E*-plane. The plane composed of the magnetic-field vector and the direction of maximum radiation is known as the *H*-plane. For the 3D plot, if the direction of maximum radiation is considered to be along the *x*-axis, then the *x*–*z*-plane is considered the *E*-plane and the *x*–*y*-plane is considered the *H*-plane.

2.1.2.2 Directivity of antennas

Two important parameters of an antenna achieved from radiation pattern are directivity and gain. These parameters are often used to identify an antenna suitable for a specific application. For example, high gain antennas as required for satellite communication [18] whereas highly directive antennas are suitable for tracking radar [15]. Directivity of an antenna can be estimated from (2.1) as the ratio of the radiation intensity from the antenna in a given direction (U) to the radiation intensity that is averaged over all directions (U_{avg}). The average radiation intensity is evaluated from total power radiated by the antenna as given in (2.1) [10].

$$\text{Directivity } (D) = \frac{U}{U_{avg}} = \frac{U}{P_{rad}/4\pi} \tag{2.1}$$

In practice, directivity of an antenna can be estimated from the beamwidth. For an antenna with one major lobe and negligible minor lobes, the directivity can be calculated using (2.2) [10] where the solid angle is approximated by the product of the beamwidths Ψ_E and Ψ_H in two principal planes (E and H plane), respectively. In (2.2) the expression is modified for HPBWs in degrees.

$$D = \frac{4\pi}{\Omega_S} = \frac{4\pi}{\Psi_E \Psi_H} = \frac{41253}{(\Psi_E)_{deg}\,(\Psi_H)_{deg}} \tag{2.2}$$

2.1.2.3 Efficiency of antennas

Antennas exhibit various types of efficiencies and the two most widely used efficiencies are radiation efficiency (η_{rad}) and aperture efficiency (η_{ap}). Radiation efficiency is a measure of the ability of an antenna to convert the RF power accepted at its input (P_{in}) into radiated power (P_{rad}). The difference between input power and radiated power is caused by various losses associated with an antenna, such as reflections caused by mismatch between the transmission line and antenna, conduction and dielectric losses [13]. Reflection from the antenna end towards input due to impedance mismatch causes reflection efficiency (η_r) and it can be estimated from input reflection coefficient Γ. Moreover, the conductor loss due to metallic components such as radiating patch, ground plane accounts for conductor efficiency (η_c) whereas dielectric loss caused by the substrate, spacers, etc., accounts for dielectric efficiency (η_d) in an antenna. Thus, total efficiency (η_T) can be expressed as [14]:

$$\left.\begin{array}{c} \eta_T = \eta_r \eta_{rad} \\ \text{where,} \\ \eta_{rad} = \eta_c \eta_d \quad \text{and} \quad \eta_r = (1 - |\Gamma|^2) \end{array}\right\} \tag{2.3}$$

On the other hand, aperture efficiency is caused by the fact that for an antenna whole of the physical aperture is not responsible for the radiation leading to two apertures: physical aperture (A_p) and effective aperture (A_e). Aperture efficiency is expressed as:

$$\eta_{ap} = \frac{A_e}{A_p} \tag{2.4}$$

In a practical situation, for antenna efficiency measurement, it is expressed in terms of loss resistance (R_{loss}) and radiation resistance (R_r) as follows [14]:

$$\eta_{rad} = \frac{P_{rad}}{P_{in}} = \frac{P_{rad}}{P_{loss} + P_{rad}} = \frac{R_r}{R_{loss} + R_r} \tag{2.5}$$

2.1.2.4 *Antenna gain*

Antenna gain is closely related to directivity of the antenna; however, with a minor difference. Gain of an antenna is related to efficiency of the antenna as well as its directional capabilities. In a given direction, gain of an antenna (G) can be estimated as the ratio of radiation intensity (U) in the desired direction, to the radiation intensity (U_i) obtained by considering that the power fed to the antenna is radiated isotropically [15]. The latter can be found from input power (P_{in}). Thus,

$$\text{Gain } (G) = \frac{U}{U_i} = \frac{U}{P_{in}/4\pi} \tag{2.6}$$

In general, gain of any antenna is measured in terms of relative gain that corresponds to the ratio of the gain in the desired direction to the gain of a reference antenna that is well known. Gain is related to the directivity as:

$$\text{Gain}(G) = \eta_{rad} \times \text{Directivity}(D) \tag{2.7}$$

Apart from this, another parameter known as Absolute Gain (G_{abs}) considers the reflection mismatches between antenna and the transmission line and thus it can be estimated from the reflection efficiency (η_r) as follows [15].

$$G_{abs} = \eta_r \times G = (1 - |\Gamma|^2)G \tag{2.8}$$

For a transceiver operating in the microwave spectrum, the receiver antenna gain (G_R) depends on effective aperture (A_e) of the antenna and wavelength (λ) at operating frequency as [13]:

$$GR = \frac{4\pi A_e}{\lambda^2} = \frac{4\pi \times \eta_{ap} A_p}{\lambda^2} \tag{2.9}$$

2.1.2.5 *Isotropic, omni-directional and directional antennas*

An antenna can be one of the three types of radiators, namely: isotropic, omni-directional and directional, depending upon the variation of radiated power level with angular (theta or phi) variation. An isotropic radiator radiates equal amounts of power in all the directions ($0 < \theta < \pi$ and $0 < \phi < 2\pi$) as shown in figure 2.4(a) and thus directivity of such antenna is $D = 1$ or 0 dB. However, an isotropic antenna exists in ideal conditions only [10]. An omni-directional antenna on the other hand radiates equal power in all directions but in a single plane, for example in the *xy*-plane as shown in figure 2.4(b) and exhibits directivity (D) slightly higher than 0 dB [13]. Omni-directional antennas, such as monopole antennas and dipole antennas,

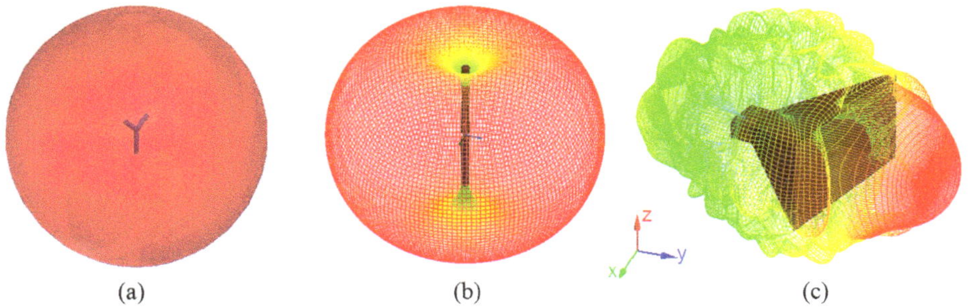

Figure 2.4. Three-dimensional radiation plot for (a) isotropic, (b) omni-directional and (c) directional antenna ('red' represents maximum power level and 'green' represents minimum power level).

find applications in Wi-Fi routers, FM radio, and base stations. A directional antenna radiates with maximum power level in a specific direction and has a major lobe with smaller HPBW [14]. Directional radiation from a horn antenna is shown in figure 2.4(c). In such antennas the difference between the major lobe and minor lobe radiation level might be more than 20 dB. Directivity of such antennas is much more than 0 dB (18 dB for the horn in figure 2.4(c)). Directional antennas are widely used in satellite, space, and astronomical applications [18].

2.1.3 Various antennas and their applications

Since the invention of the monopole antenna by G Marconi [11] in 1895 and the construction of first horn antenna by Sir J C Bose in the year 1897 [19], there has been extensive research in the field of antenna engineering. Different types of antennas have been developed for various applications over time [16]. Some of them include modification of the previously developed antennas, whereas novel antenna designs have also been proposed. An overview of some of the conventional as well as some modern antennas [17] is given in this section.

2.1.3.1 Wire antenna

Wire antennas are the simplest in terms of design complexity, cost effective and at the same time exhibit versatile applications. Dipole antennas, monopole antennas, and helix and loop antennas are some of the widely used wire antennas. A dipole of length l consists of two conductors each of length $l/2$ with a small gap. An infinitesimal dipole has a length of $l \ll \lambda$ (typically $l < \lambda/50$) whereas a small dipole has a length between $\lambda/50 < l < \lambda/10$ [10]. However dipoles of length comparable to the wavelength especially half-wave dipole with $l = \lambda/2$ is frequently used for communication applications that require omni-directional radiation.

Radiation intensity of a half-wave dipole antenna can be found from its field components and can be expressed in terms of current on the antenna I_0 and angle θ as [13]:

$$U = \eta \frac{| I_0 |^2}{8\pi^2} \sin^3 \theta \tag{2.10}$$

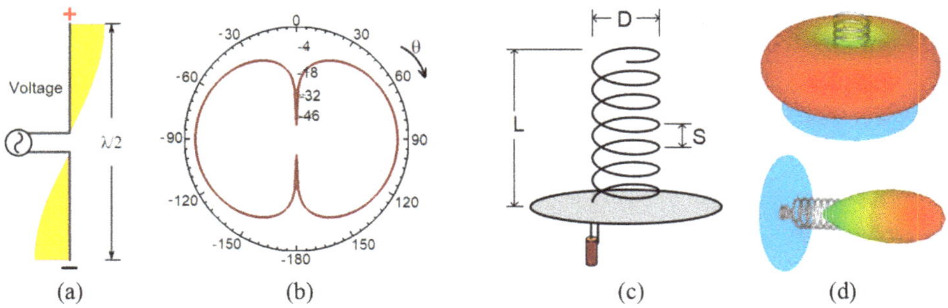

Figure 2.5. (a) A half-wave dipole antenna. (b) Two-dimensional radiation plot of dipole antenna. (c) A helical antenna with ground plane. (d) Two modes of radiation from the helical antenna.

where η is impedance of the free space (120π ohm). A half-wave dipole with voltage distribution and its two-dimensional radiation plot for variation of θ are shown in figures 2.5(a) and (b), respectively. It is apparent that the plot has a null along $\theta = 0°$ and 180° and the antenna exhibits omni-directional radiation in the plane of $\theta = 90°$.

A dipole antenna has a radiation resistance (R_{rad}) of 73 ohms [10], which is close to characteristic impedances of most of the transmission lines, i.e., 50 or 75 ohm that makes it easier to match to the line. Maximum directivity of a dipole antenna is found for maximum radiation intensity ($U = U_{max}$ in equation (2.1)) at $\theta = 90°$ to be around 1.643. Directivity close to 1 accounts for the omni-directional radiation from the antenna, as also shown in figure 2.4(b).

Another configuration of wire antennas with applications in microwave communication is the helical antenna, which is formed by winding a conducting wire in the form of a screw thread, as shown in figure 2.5(c). It is a modified version of a quarter wavelength wire monopole antenna and thus contains a ground plane with the dimension on the order of $3\lambda/4$ below the helix [20]. The helix is generally connected to the centre conductor of a coaxial transmission line. A helical antenna can have N turns and with a spacing of S between each turn, its total length becomes L = NS. Total length (L_T) of the wire is related to S and circumference of the helix $C = \pi D$ (D being the diameter) as $L_T = N\sqrt{S^2 + C^2}$ [20]. Another important dimension is the pitch angle α between a tangent to the helix and a plane perpendicular to the helix, and is defined by [13]:

$$\alpha = \tan^{-1}\frac{S}{\pi D} = \tan^{-1}\frac{S}{C} \tag{2.11}$$

Radiation characteristics of the antenna can be controlled by varying its dimensions compared to the wavelength. The pitch angle and size of the conducting wire determine the input impedance of the antenna. Helical antennas can operate in two modes depending on its radiation pattern: normal mode and axial mode [20]. The normal mode has maximum radiation in a plane perpendicular to the axis and null along the axis, similar to the dipole antenna radiation as shown in figure 2.5(d) at the top. Normal mode can be achieved for total length $L_T \ll \lambda_0$ and pitch angle α close to zero. Besides, for the axial mode maximum radiation occurs along the axis of the helix similar to that of an end-fire radiation as shown in figure 2.5(d) at the bottom. This mode can be generated by keeping diameter D and spacing S as large fractions of the wavelength λ_0 such that:

$$\frac{3\lambda_0}{4} < \pi D < \frac{4\lambda_0}{3} \text{ and } S \approx \frac{\lambda_0}{4}$$

Helical antennas are primarily elliptically polarized antennas but can receive signals from a rotating linearly polarized antenna. Thus they can be useful while placed on the ground for space telemetry applications of satellites, space probes, and ballistic missiles [10, 18]. Moreover, at different frequencies helical antennas can also be used as circularly polarized antennas.

2.1.3.2 Array antenna

The antennas mentioned in the previous section are single element antennas and produce a radiation pattern with wide beamwidth resulting in smaller directivity and gain. However in applications involving long range communication such as satellite communication, mobile communication, etc., antennas with high directivity are required [14]. An array of the individual antenna elements placed in appropriate electrical and geometrical configuration can serve this purpose. In general the elements of an array antenna are identical.

An array of two small elements with a spacing '*d*' is shown in figure 2.6(a). The elements can be a small dipole or monopoles. The resultant radiated field at an observation point P of the array can be estimated by vector addition of the fields radiated by individual antenna elements. The current in each element is assumed to be the same as that of the isolated element. Assuming there is no coupling between the elements, the calculation of total field of the array shows that the total field is related to the field of a single element positioned at origin O as [10]:

$$(\text{Electric field})_{\text{array}} = (\text{Electric field})_{\text{single element}} \times \text{Array factor(AF)}$$

Here the array factor for the two-element array of constant amplitude and a difference in phase excitation of β between the elements in the far field is given by [13]:

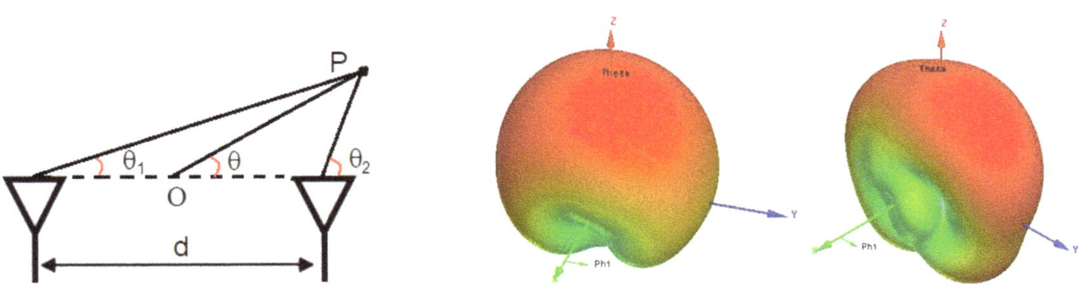

(a) (b) (c)

Figure 2.6. (a) An array antenna with two elements. (b) 3D radiation plot of single element. (c) 3D radiation plot of the two element array.

$$\mathrm{AF} = 2\cos\left[\frac{1}{2}\left(kd\cos\theta + \beta\right)\right] \tag{2.12}$$

Thus, by varying the array element spacing d and/or the phase difference β between the elements, the array factor and the total field of the array can be controlled. The radiation pattern of an array results from what is known as pattern multiplication for arrays of identical elements. Figure 2.6(b) shows the 3D radiation plot of a single element, which after multiplied by the Array Factor results in the radiation pattern as shown in figure 2.6(c) for a two-element array [14]. This plot is shown for a linear array with the two elements placed in a single row along the x-axis. It is apparent that the radiation becomes directive only along the x-axis. A similar procedure is applicable for the estimation of total field or radiation pattern of array of more than two elements. In order to enhance the directivity along both the x- and y-axis, a two-dimensional array is needed.

2.1.3.3 Printed antenna

Printed antennas present an excellent solution and take an important role in the development of the new generation of wireless and mobile communication systems in the microwave regime. By the name 'printed' it refers to an antenna fabricated using a photolithographic process on a printed circuit board (PCB) and thus can be easily integrated into portable devices, wireless sensors, and mobile equipment [13, 14]. Printed antennas are popular for their light weight, ease of fabrication, low cost, wide range of operating frequencies, greater form-factor capability for flexible and disposable applications.

The microstrip patch antenna is the fundamental among various types of printed antennas. It consists of a conducting patch of any shape placed above a dielectric substrate with a ground plane on the other side, as shown in figure 2.7(a). The patch can be given excitation using a microstrip line feed, as shown, or one of the other feeding methods such as coaxial probe feed, aperture coupled feed, and proximity coupled feed [21]. The electric field is usually radiated along the patch width (W) whereas the resonating frequency is decided by the length (L). The length depends on guided wavelength and the fringing effect caused by the electric field that in turn

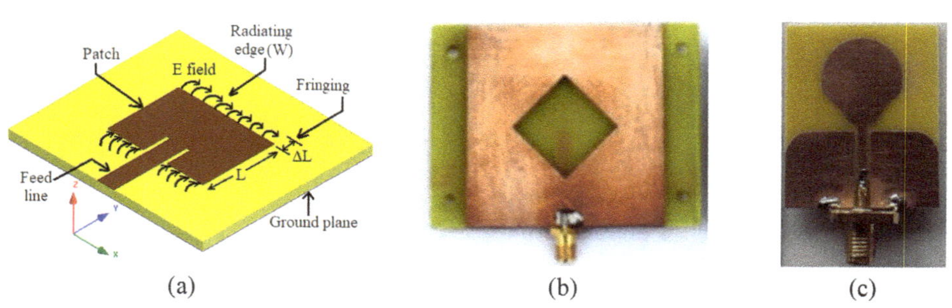

Figure 2.7. (a) A microstrip line fed rectangular patch antenna, (b) wide slot antenna with microstrip line fed on the other side [22], and (c) CPW fed patch antenna with ultra-wideband response. Reproduced from [23]. CC BY 4.0.

increases the effective length of the patch ($L_{eff} = L + 2\Delta L$) [10]. However, the patch antenna suffers from surface wave loss and narrow bandwidth affecting antenna efficiency.

Various other configurations of printed antenna are available with enhanced characteristics. Slot antennas, as shown in figure 2.7(b), have a radiating slot in the ground plane that is fed via a microstrip line on the other side of the substrate [22]. The slot here functions as a magnetic dipole. Depending upon the enhanced slot width, impedance bandwidth (−10 dB) of the antenna increases leading to a wideband response. An ultra-wideband response (3.1–10.6 GHz) can be achieved by coplanar waveguide structure (CPW) where the ground plane and radiating patch are on the same side of the substrate as shown in figure 2.7(c) [23]. The impedance matching is further improved by using tapering on the ground plane as shown. Printed UWB antenna has a wide range of applications including radar imaging, tracking, wireless sensor networks, and ground penetrating radar (GPR) [23].

2.2 Fundamentals of frequency selective surfaces

Besides antennas, filters are also known to be a fundamental part of most of the electronic as well as radio frequency (RF) circuits in the microwave regime. Filters are incorporated in a communication system for controlling frequency content in the signal at the output, thereby discarding the noise and interference [1]. Traditional filters are categorized as low pass, high pass, band pass and band stop filters. Frequency selective surfaces (FSSs) are the wireless counterpart of traditional filters operating at microwave spectrum and beyond [24, 25]. It is a two-dimensional periodic array of metallic patches on a dielectric slab or array of perforations on a metallic screen above a dielectric slab, without any ground plane on the other side of the substrate [24]. Just like RF filters, FSSs can be any of the four types depending on its geometry and dimension; however, band pass and band stop FSS structures are used primarily. FSSs, once exposed to the electromagnetic radiation, pass certain frequency components, whereas they block and reflect back others.

2.2.1 Geometry of FSS

Since the inception in the early 1960s, FSS structures developed by researchers have been extensively used in defence applications [24, 26], however Marconi and Franklin are the early pioneers in this field. They incorporated half-wavelength wire sections in a parabolic reflector in 1919 [25]. In the earlier days, FSSs were primarily used as covers of antennas or other secured objects, also known as radomes, for protecting them from environmental hazards and shielding them from enemy radars, by reducing radar cross section (RCS) of the objects (figure 2.8) [31].

According to its definition, FSS can be either an array of patches or apertures. A patch-type FSS array with square shaped elements is shown in panel (a) with its transmission response. The region shown with dashed lines is known as the 'unit cell'. It basically acts as a low-pass filter with a capacitance between the consecutive patches. In contrast the aperture-type FSS composed of square shaped grids, as shown in panel (b), acts as a high-pass filter, with inductance along the grid, as

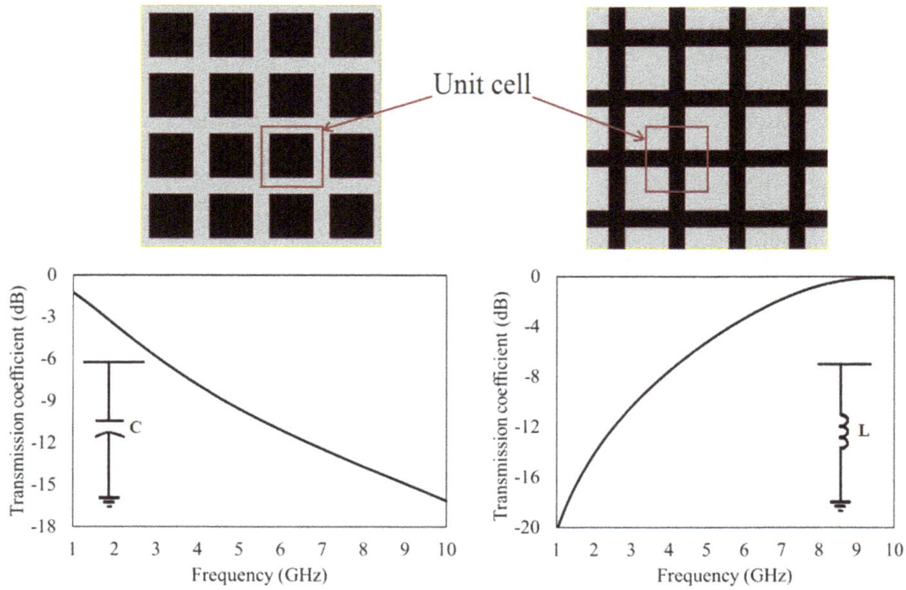

(a) Patch type FSS (b) Aperture type FSS

Figure 2.8. Patch and aperture type FSS arrays along with their transmission response.

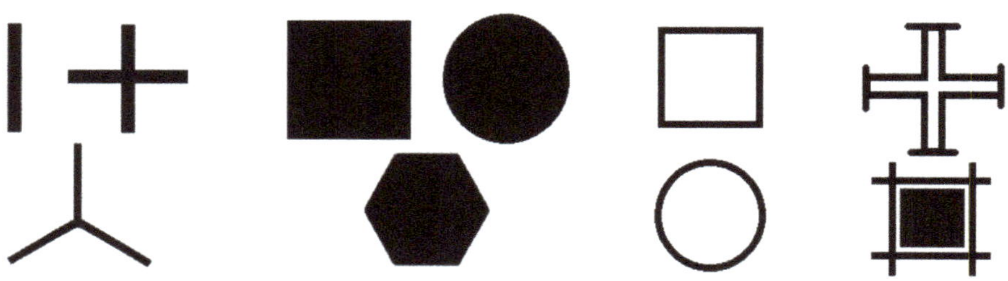

(a) Centre connected (b) Solid patch (c) Loop shaped (d) Hybrid

Figure 2.9. Various shapes of unit cells of the patch type frequency selective surfaces. [24] John Wiley & Sons. [Copyright © 2005 John Wiley & Sons, Inc. All rights reserved.]

confirmed by the transmission response. Similar to traditional filters, in case of FSS, bandpass and bandstop responses can be achieved by cascading multiple FSS layers with low-pass and high-pass transmission responses [26–28].

Since its invention, a variety of unit cells have been developed and used for different applications, as shown in figure 2.9 [24]. Some of them, as can be seen, are dipole, crossed dipole, and tripole (figure 2.9(a)), solid metallic patches in the shape of circle and hexagon (figure 2.9(b)), loop shaped elements such as square or circular ring (figure 2.9(c)), and hybrid elements such as combination of crossed dipole with loop or crossed dipole with solid patch (figure 2.9(d)) [24, 26]. For any element of patch-type FSS the corresponding aperture-type FSS can be built just by taking the complement, that is by replacing the metal with void and the surrounding area replaced by metal according to the principle of Babinet [30]. Detailed classification of the elements along with their dimensions for FSS design is also presented in the subsequent section.

2.2.2 Principle of operation

The principle of operation of FSS can be understood from figure 2.10, where a dipole shaped patch-type FSS is considered with each dipole having half-wavelength dimensions. The spacing between consecutive dipoles is much less than the wavelength. Plane waves from a source are incident on the FSS at an angle of 90° (normal incidence). The patches being made of metal, the electric fields exert a force on the electrons and cause them to oscillate as can be seen. A portion of energy from the waves is converted to kinetic energy of the electrons and they remain in the oscillating state, allowing a surface current to flow in the patches, so an electric field is developed between the consecutive patches. Oscillating current in the patches can be modeled by an inductor (L) whereas the electric field between the patches can be modeled by a capacitor (C). The series combination of L and C forms a bandstop filter [29].

If the wavelength corresponding to the frequency of plane wave matches resonating length of the dipoles ($\lambda/2$) then a large portion of energy gets absorbed by the electrons and resonance occurs. Under this condition oscillating current on the patches acts as a source of electromagnetic waves and the resultant waves get reflected back towards the source, causing poor transmission through the FSS as evident from figure 2.10(a). The reflection gives a bandstop response with the resonating frequency corresponding to L and C values, depends on the patch dimension (for L) and spacing between the consecutive patches (for C).

On the other hand, if the waves fall on an FSS that has a dimension different (here smaller) from the previous one, as shown in figure 2.10(b), then due to mismatch, resonance does not occur. Accordingly, electrons absorb much less energy and due to the small amount of current flow, most of the waves get transmitted through the FSS causing poor reflection. So the FSS acts like a transparent screen to the waves at

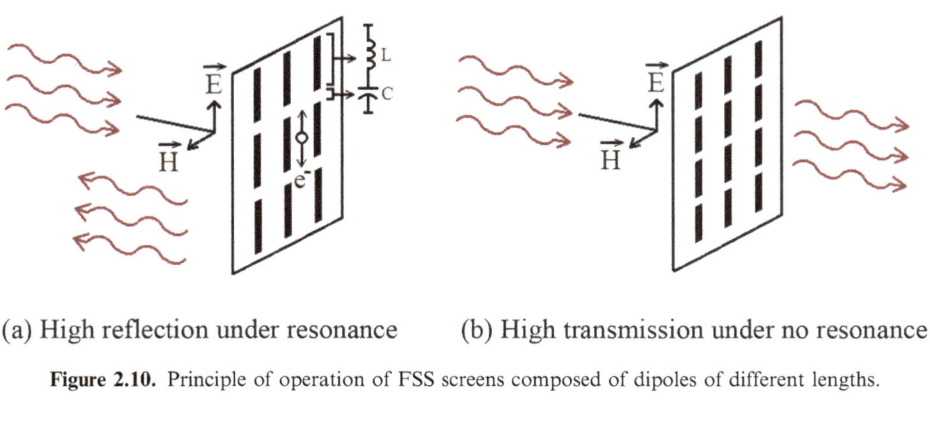

(a) High reflection under resonance (b) High transmission under no resonance

Figure 2.10. Principle of operation of FSS screens composed of dipoles of different lengths.

frequencies outside the stop band. Similarly in the case of aperture-type FSSs, a parallel LC resonator is formed and signals passes through the FSS with negligible insertion loss at a certain frequency band (pass band) and incur reflection at all other frequencies.

2.2.3 Conventional frequency selective surfaces

The operating principle of a frequency selective surface with the incidence of a plane wave, as discussed previously, shows that the FSS basically operates at resonance. As a result, element dimensions are chosen so as to resonate at the frequency of operation. However, apart from dimension, shape of the FSS elements is also important, as for the same frequency, different shapes need different dimensions, and depending on the application, shapes with lower profile should be chosen [24, 26–28]. Based on element shape of the unit cell, in general three distinct classes of FSS, as was introduced in the section 2.2.1, can be discussed as follows:

(i) The center connected elements
(ii) The loop shaped elements
(iii) Solid plate type elements

For the purpose of analysis, various element shapes have been simulated, and transmission and reflection characteristics are presented in this section for each class of elements. Simulations are performed using ANSYS HFSS. Arlon AD270 substrate of 1.2 mm thickness, dielectric constant of $\varepsilon_r = 2.7$ and loss tangent of 0.002 is chosen primarily for the designs.

2.2.3.1 The center-connected elements

Center-connected elements are constructed by connecting two or more monopole strips at the center, and accordingly various shapes can be achieved, such as a dipole with two legs, a tripole with three legs, crossed dipoles with four legs and so on [28]. Various examples of such elements are shown in figure 2.11 [24]. An array of crossed dipoles is shown in figure 2.12 along with its frequency response. The dimension of the unit cell is also included in the plot for better understanding. In this case the crossed dipoles are loaded with small, orthogonal sections at the ends, and this shape is popularly known as the Jerusalem cross, one of the oldest building blocks in FSS design [32]. This is the modified version of the crossed dipole shaped element without stubs or end loading, as depicted in the last row of figure 2.11.

The unit cells can be placed in a compact manner with reduced inter-element spacing, thereby enhancing the bandwidth [33] unlike the arrangement in figure 2.12.

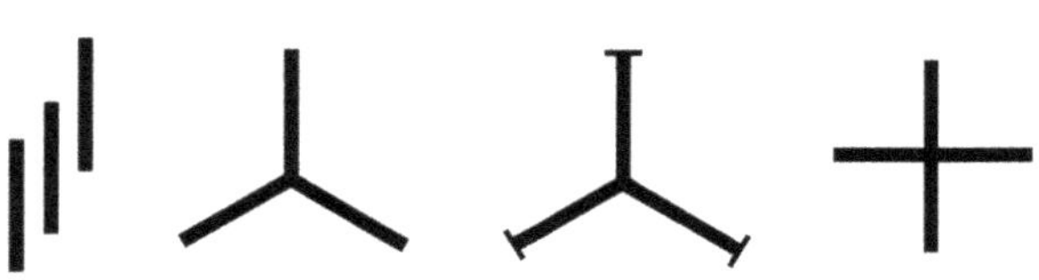

Figure 2.11. Various shapes of center-connected elements used in FSS. [24] John Wiley & Sons. [Copyright © 2005 John Wiley & Sons, Inc. All rights reserved.]

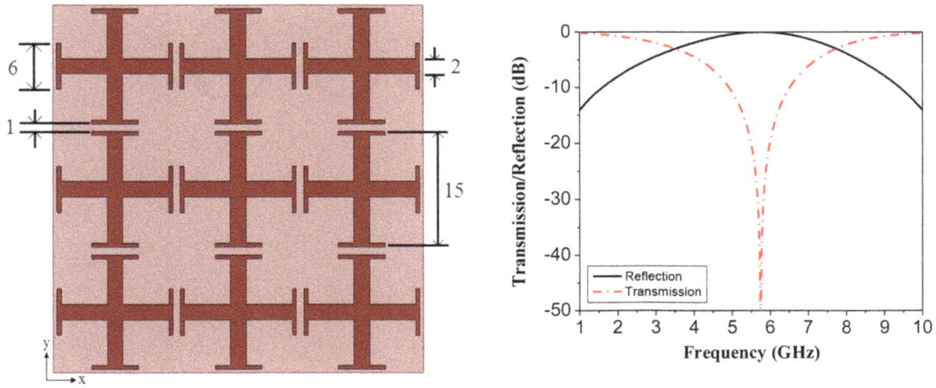

Figure 2.12. Periodic array of Jerusalem cross-shaped elements (all dimensions are in mm) and transmission/reflection response of the FSS.

Advantages of the Jerusalem cross are better angular stability compared to the tripole elements [28], and ease of frequency tuning by varying the end loading strip length, that changes the capacitance between each element. Although the unit cell dimension remains the same for the Jerusalem cross FSS, there is a downshift in resonance due to the end loading. Unlike dipole and tripole arrays, the crossed dipole array is independent of horizontal and vertical polarization due to the symmetric nature [24]. Polarization independency of the FSS is extremely useful in applications involving dual-polarized antennas, where the same FSS can be used for both the polarizations.

2.2.3.2 The loop shaped elements

The second class of elements is in the shape of a loop, which can be a square loop, circular loop, or hexagonal loop [24]. Elements with multiple folded legs also come in this category, such as three- or four-legged elements. Some of the shapes are shown in figure 2.13. Three- and four-legged elements are the first and second examples in the figure.

An array of simple square loop-shaped patches is shown in the figure 2.14 along with its frequency response. Its resonance mechanism can be easily understood from the figure. A square loop can be thought of being composed of two folded dipoles connected to each other diagonally. In that case each side of the folded dipole has a length of $\lambda/4$, leading to a quarter wavelength dimension of the square loop [24]. In this example, for a length of 14 mm (equivalent to $\lambda_g/4$) of the loop, it should resonate near 4.2 GHz for the similar dielectric material used before, and it can be verified from the transmission response in figure 2.14. The shift can be adjusted by slightly varying the periodicity or spacing between each element. The spacing should be kept much less than the wavelength. As discussed before, larger spacing causes early onset of the grating lobes [33], and also lead to smaller bandwidth. For oblique incidence of the plane wave, smaller spacing helps in avoiding the harmonics. Hexagonal elements as shown in figure 2.13 exhibit better harmonic performance. It has a better angular stability and can be used in scanning applications [28].

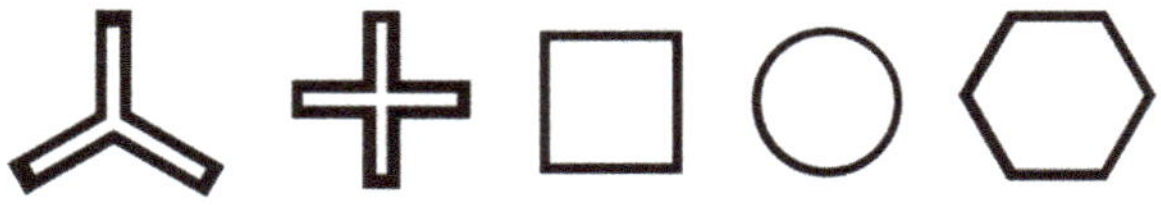

Figure 2.13. Various types of loop-shaped elements used in FSS design. Reproduced from [28]. CC BY 4.0.

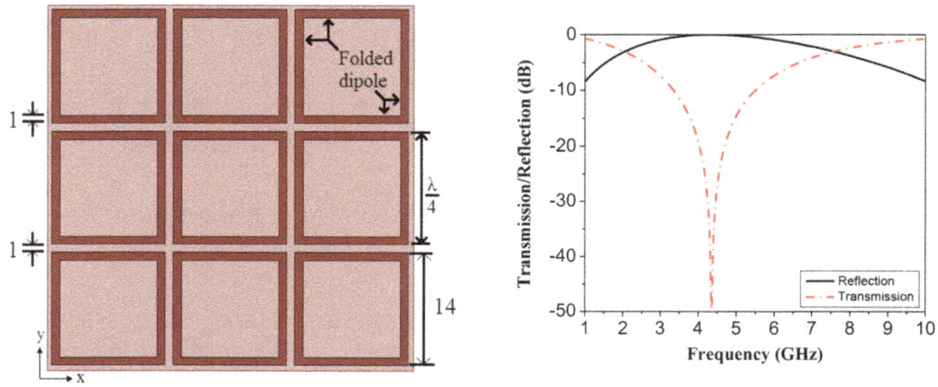

Figure 2.14. Periodic array of square loop-shaped elements (all dimensions are in mm) and transmission/reflection response of the FSS.

2.2.3.3 Solid plate type elements

Solid plate type elements can take the shape of a square, rectangle, circle, or hexagon, as shown in figure 2.15. The elements in this class can take one of the two forms: (1) array of metallic patches and (2) array of apertures or perforations within a metallic screen [26]. These two are complementary structures and are discussed here.

The unit cell dimension of patch or aperture is generally $\lambda/2$, leading to poor angular stability and early onset of the grating lobes [33]. However they can be used with certain modifications such as with the introduction of slots in the patch. These elements also suffer from shift in resonance and change in bandwidth that limits the use of these elements in many applications. An array of square patches and that of square grids or accordingly square apertures on a metallic screen were studied in the previous section and was depicted in figure 2.8. It was observed that they do not resonate and rather act as low pass or high pass filter. However resonance could be observed for these elements at a much higher frequency (beyond the range shown), that could be shifted down by increasing the inter-element spacing. But this would again lead to poor angular stability as mentioned in the previous section. However a combination of these elements may result in highly miniaturized unit cells.

2.2.3.4 Complementary FSS structures

In order to investigate complementary FSS structures crossed dipole shaped element are considered. An array of patches and that of apertures are shown in figures 2.16 (a) and (b), respectively, along with their transmission response. These complementary structures follow Babinet's principle, which states that complementary structures or elements of the same dimensions exhibit identical resonance [30].

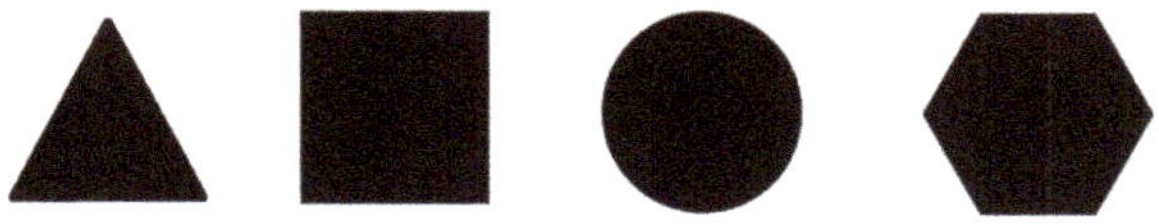

Figure 2.15. Various shapes of solid plate type elements used in FSS design.

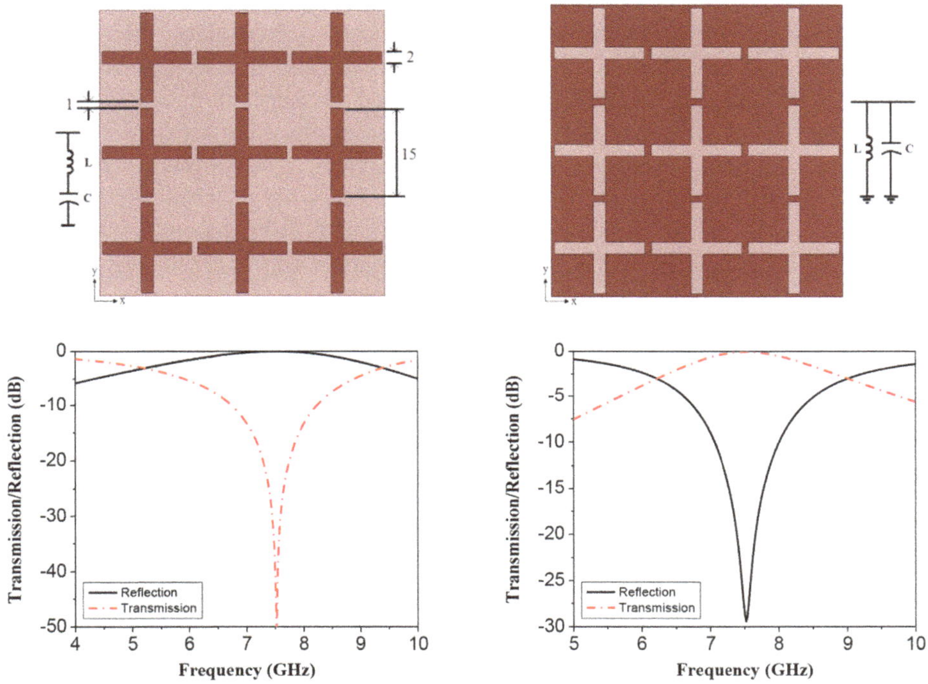

(a) Array of crossed dipole patches (b) Array of crossed dipole apertures

Figure 2.16. Comparison between an array of crossed dipole patches and that of apertures, along with their transmission and reflection response (dimensions are in mm).

However, there are certain conditions for this principle to be satisfied such as: the metallic layer must be thin (around $\lambda/1000$), and substrate thickness must be only a fraction of wavelength [26]. Thicker dielectric material and the use of multiple layers in designing the FSS unit cell vary the frequency response. As evident from the transmission response in figure 2.16 the crossed dipole patch array exhibits bandstop response at 7.5 GHz whereas complementary structure of the same element, i.e., array of crossed dipole shaped apertures shows bandpass response at 7.5 GHz only. Thus it is apparent that the complementary structures obey Babinet's principle [30, 33].

2.2.4 Applications of frequency selective surfaces

Since its introduction, frequency selective surfaces have been used in several applications, however the field or domain has changed from time to time. Earlier

FSS structures were extensively used as sub-reflectors in Cassegrain antennas for satellite communication [34], and radomes as a cover for secured objects like antennas in defence systems [35], for reducing their radar cross section (RCS) [31]. Later more fields were evolved, such as stealth applications in aircrafts [36], polarizers [37], microwave absorbers [38], EMI (electro-magnetic interference) protection, microwave ovens, and secured indoor communication [39].

FSS structures based on both planar and conformal surfaces have been utilized in multi-frequency reflector antennas for the purpose of frequency division multiplexing. Such antennas find their applications in science investigations, and data communication links in satellite communication. Cassegrain antennas generally need dichroic-based sub-reflectors [34] as in such antennas there are two feeds mostly horn antennas, namely primary feed and secondary feed. As shown in figure 2.17, the radiation at frequency f_1 from primary feed reaches the parabolic reflector after getting reflected by the FSS, whereas the radiation at f_2 from secondary feed passes through the FSS that is not reflective at f_2, and reaches the parabolic reflector. The dichroic used here can also be bandpass in nature [34].

Radomes are used to cover antennas in order to protect them against environmental hazards. However instead of dielectric radomes, radomes based on multi-layered FSSs are being mainly used in defence applications such as radar and missiles for preventing the antennas and other secured objects from getting detected by the enemy radar, and this is performed by reducing the radar cross section of the objects [35], the ability of a target of being detected. Radomes can be of various shapes such as hemispherical, conical, and cylindrical [40]. The unit cell of the FSS used for constructing the radome can be triangular or hexagonal. A hemispherical radome covering an antenna is shown in figure 2.18(a), and in this case triangular shaped unit cells have been used. Radomes are used also for radio telescopes, and naval and airborne antennas [35]. FSS layers mounted on an aircraft nose can be seen from figure 2.18(b) where planar FSS is used that is transparent in the operating band of the antenna array and acts like a perfect electric conductor outside the band. The in-band signals smoothly reach the antenna, whereas the out-of-band signals get

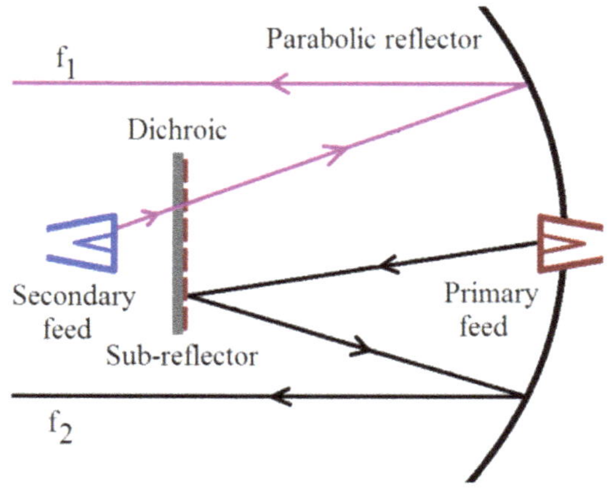

Figure 2.17. Dichroic sub-reflector in Cassegrain antenna for frequency division multiplexing.

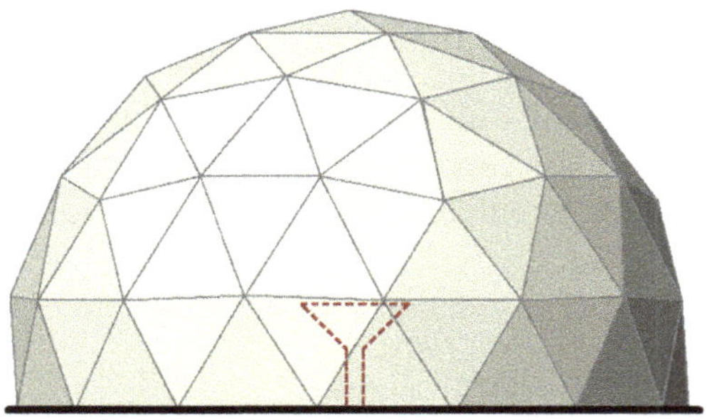

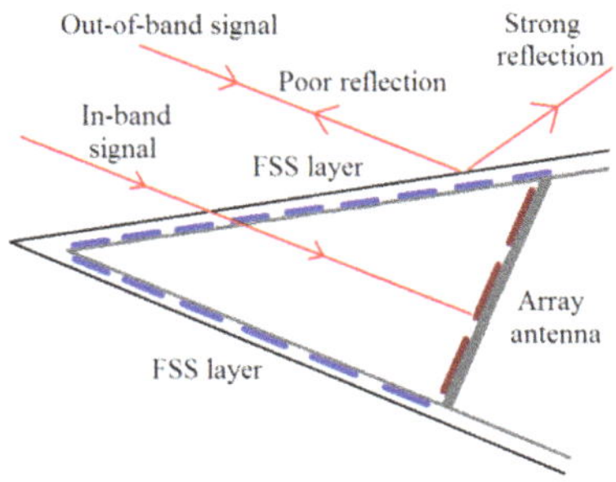

Figure 2.18. (a) Hemispherical radome with triangular unit cells, covering an antenna (shown in dotted line). (b) FSS mounted on the nose of an aircraft for RCS reduction.

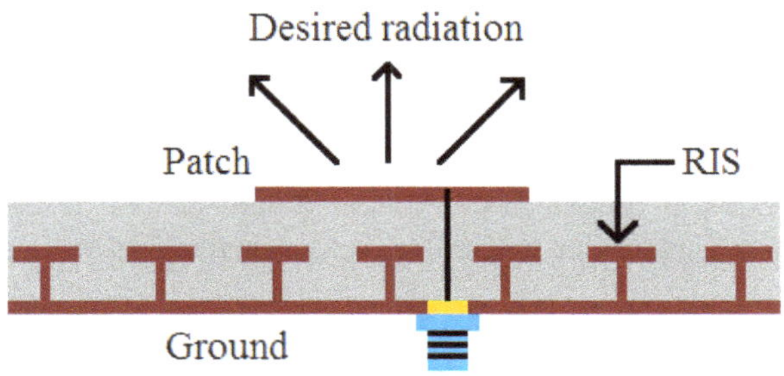

Figure 2.19. Application of FSS as a reactive impedance surface (RIS) in a patch antenna.

reflected by the FSS mostly towards a direction other than the direction of incidence (bi-static).

Apart from traditional applications, FSS can be used as a polarizer, a diffraction grating, from which the fields that have polarization similar to the grating are reflected, while the fields that have perpendicular polarization are transmitted [37]. Depending on the shape of FSS unit cell, polarizer can be of few types. Meander line FSS screens are used to convert horizontal polarization to vertical one. In satellite communication circular polarization is extremely useful, and FSS-based metasurfaces of specific geometry can be used as a superstrate above a linearly polarized antenna to achieve circularly polarized radiation with either RHCP (right hand circular polarization) or LHCP (left hand circular polarization) [41].

Frequency selective surfaces, when properly designed to exhibit a certain reflection phase and magnitude, can be used as a reactive impedance surface (RIS) [42]. FSS structures, when placed above a ground plane, with or without via connecting the FSS and ground, form a RIS as shown in figure 2.19. It can be used for miniaturizing the antenna dimension and improving its impedance matching [43]. FSSs can also be used as an artificial magnetic conductor (AMC) [44] that has a construction similar to RIS and is used for different applications. A magnetic conductor cannot support electric current unlike PEC and suppresses the propagation of surface waves inside the antenna substrate. So AMC is placed between the patch and ground plane to enhance the bandwidth as well as to improve the radiation efficiency. It is extremely useful in increasing the gain of some antennas that have poor gain [45], such as electrically small antennas, however, the enhancement is achieved over a narrow bandwidth. FSSs can be incorporated in CPW fed ultra wideband antennas (UWB) as a reflector to enhance their broadside radiation in terms of gain [46, 47].

References

[1] Tse D and Viswanath P 2005 *Fundamentals of Wireless Communication* (Cambridge: Cambridge University Press)

[2] Gow G and Smith R 2006 *Mobile and Wireless Communications: An Introduction* (London: McGraw-Hill Education)

[3] Haykin S S and Moher M 2011 *Modern Wireless Communications* (Pallavaram: Pearson Education India)

[4] Arya K V, Bhadoria R S and Chaudhari N S 2018 *Emerging Wireless Communication and Network Technologies* (Berlin: Springer)

[5] Qiu L, Zhao F, Xiao K, Chai S L and Mao J J 2012 Transmit–receive isolation improvement of antenna arrays by using EBG structures *IEEE Antennas Wirel. Propag. Lett.* **11** 93–6

[6] Rahimi M, Zarrabi F B, Ahmadian R, Mansouri Z and Keshtkar A 2014 Miniaturization of antenna for wireless application with difference metamaterial structures *Prog. Electromagn. Res.* **145** 19–29

[7] Kurra L, Abegaonkar M P, Basu A and Koul S K 2016 FSS properties of a uniplanar EBG and its application in directivity enhancement of a microstrip antenna *IEEE Antennas Wirel. Propag. Lett.* **15** 1606–9

[8] Monebhurrun V 2020 Revision of IEEE standard 145-2013: IEEE standard for definitions of terms for antennas [stand on standards] *IEEE Antennas Propag. Mag.* **62** 117

[9] Kraus J D 1988 Heinrich Hertz-theorist and experimenter *IEEE Trans. Microw. Theory Tech.* **36** 824–9

[10] Balanis C A 2016 *Antenna Theory: Analysis and Design* (New York: Wiley)

[11] Ramsay J 1981 Highlights of antenna history *IEEE Commun. Mag.* **19** 4–8

[12] Belrose J S 2001 A radioscientist's reaction to marconi's first transatlantic wireless experiment-revisited *IEEE Antennas and Propagation Society International Symposium. 2001 Digest. Held in Conjunction with: USNC/URSI National Radio Science Meeting (Cat. No. 01CH37229* **Vol 1** pp 22–5 (Piscataway, NJ: IEEE)

[13] Kraus J D, Marhefka R J and Khan Ahmad S 2006 *Antennas and Wave Propagation* (New York: Tata McGraw-Hill Education)

[14] Elliot R S 2006 *Antenna Theory and Design* (New York: Wiley)

[15] Saunders S R and Aragón-Zavala A 2007 *Antennas and Propagation for Wireless Communication Systems* (New York: Wiley)

[16] Huang Y 2021 *Antennas: From Theory to Practice* (New York: Wiley)

[17] Janaswamy R 2001 *Radiowave Propagation and Smart Antennas for Wireless Communications* (Berlin: Springer Science & Business Media)

[18] Rahmat-Samii Y and Densmore A C 2014 Technology trends and challenges of antennas for satellite communication systems *IEEE Trans. Antennas Propag.* **63** 1191–204

[19] Sen A K 1997 Sir JC Bose and radio science *1997 IEEE MTT-S International Microwave Symposium Digest* **vol 2** pp 557–60 (Piscataway, NJ: IEEE)

[20] Djordfevic A R, Zajic A G, Ilic M M and Stuber G L 2006 Optimization of helical antennas [antenna designer's notebook] *IEEE Antennas Propag. Mag.* **48** 107–15

[21] Guha D and Antar Y M (ed) 2011 *Microstrip and Printed Antennas: New Trends, Techniques and Applications* (New York: Wiley)

[22] Chatterjee A and Parui S K 2015 Gain enhancement of a wide slot antenna using a second-order bandpass frequency selective surface *Radioengineering* **24** 455–61

[23] Kundu S and Chatterjee A 2022 A compact super wideband antenna with stable and improved radiation using super wideband frequency selective surface *AEU-Int. J. Electron. Commun.* **150** 154200

[24] Wu T K 1995 *Frequency Selective Surfaces* (New York: Wiley)

[25] Marconi G and Franklin C S 1919 Inventors; Marconi Wireless Telegraph Co America, Assignee. Reflector for use in wireless telegraphy and telephony. United States Patent US 1,301,473

[26] Munk B A 2005 *Frequency Selective Surfaces: Theory and Design* (New York: Wiley)

[27] Panwar R and Lee J R 2017 Progress in frequency selective surface-based smart electromagnetic structures: a critical review *Aerosp. Sci. Technol.* **66** 216–34

[28] Anwar R S, Mao L and Ning H 2018 Frequency selective surfaces: a review *Appl. Sci.* **8** 1689

[29] Costa F, Monorchio A and Manara G An overview of equivalent circuit modeling techniques of frequency selective surfaces and metasurfaces *Appl. Comput. Electromagn. Soc. J.* **2014** 960–76

[30] Liu X, Lu R, Zhu S, Xu Z, Chen X and Zhang A 2018 Analysis of complementary metasurfaces based on the babinet principle *IEEE Microw. Wirel. Compon. Lett.* **29** 8–10

[31] Xie D, Liu X, Guo H, Yang X, Liu C and Zhu L 2016 A wideband absorber with a multiresonant gridded-square FSS for antenna RCS reduction *IEEE Antennas Wirel. Propag. Lett.* **16** 629–32

[32] Silva H V, Silva C P, de Oliveira M R, de Oliveira E M, de Melo M T, de Sousa T R and Gomes A 2017 Multiband FSS with fractal characteristic based on Jerusalem cross geometry *J. Microw. Optoelectron. Electromagn. Appl* **16** 932–41

[33] Bayatpur F 2009 Metamaterial-inspired frequency-selective surfaces Doctoral Dissertation, University of Michigan

[34] Han M, He M, Sun H J, Zhao G Q and Cheng H 2013 Analysis of cassegrain antenna by using a dichroic sub-reflector *2013 IEEE Int. Conf. on Microwave Technology and Computational Electromagnetics* pp 58–60 (Piscataway, NJ: IEEE)

[35] Lee K W, Jeong Y R, Hong I P, Lee M G, Chun H J and Yook J G 2011 Simple design method of FSS radome analysis using equivalent circuit model *IEICE Electron. Express* **8** 2002–9

[36] Narayan S, Sangeetha B, Sruthi T V, Shambulingappa V and Nair R U 2017 Design of low observable antenna using active hybrid-element FSS structure for stealth applications *AEU-Int. J. Electron. Commun.* **80** 137–43

[37] Littman N M, O'keefe S G, Galehdar A, Espinosa H G and Thiel D V 2021 Ultra-thin broadband transmission FSS for linear polarization rotation *IEEE Access* **9** 127335–42

[38] Wei J *et al* 2019 Flexible design and realization of wideband microwave absorber with double-layered resistor loaded FSS *J. Phys.* D **52** 185101

[39] Roberts J, Ford K L and Rigelsford J M 2015 Secure electromagnetic buildings using slow phase-switching frequency-selective surfaces *IEEE Trans. Antennas Propag.* **64** 251–61

[40] Zhou H, Qu S, Lin B, Wang J, Ma H, Xu Z, Peng W and Bai P 2012 Filter-antenna consisting of conical FSS radome and monopole antenna *IEEE Trans. Antennas Propag.* **60** 3040–5

[41] Shukoor M A and Dey S 2023 Novel miniaturized arrow linked figure of eight square loop FSS-based multiband linear-circular and linear-cross reflective microwave polarizer *Int. J. Microw. Wireless Technol.* **15** 591–9

[42] Guthi S and Damera V 2021 High gain and wideband circularly polarized S-shaped patch antenna with reactive impedance surface and frequency-selective surface configuration for Wi-Fi and Wi-Max applications *Int. J. RF Microw. Comput.-Aided Eng* **31** e22865

[43] Mahendran K and Gayathri R 2021 Performance improved triangular multi band antenna using reactive impedance substrate and frequency selective surface *Inf. Technol. Ind* **9** 486–92

[44] Boukern D, Bouacha A, Aissaoui D, Belazzoug M and Denidni T A 2021 High-gain cavity antenna combining AMC-reflector and FSS superstrate technique *Int. J. RF Microw. Comput.-Aided Eng* **31** e22674

[45] Azemi S N, Beson M R, Amir A, Azhari M S and Jiunn N K 2021 Gain enhancement of CPW antenna for IoT applications using FSS with miniaturize unit cell *Phys.: Conf. Ser.* **1962** 012052

[46] Kundu S and Chatterjee A 2021 Sharp triple-notched ultra wideband antenna with gain augmentation using FSS for ground penetrating radar *Wirel. Pers. Commun.* **117** 1399–418

[47] Kundu S and Chatterjee A 2022 A new compact ultra-wideband frequency selective surface with angular stability and polarization independence for antenna radiation enhancement in microwave imaging application *AEU-Int. J. Electron. Commun.* **155** 154351

Chapter 3

Design of metamaterial-based antenna and FSS for microwave and terahertz applications

Metamaterials are beyond conventional and naturally available materials with unusual but useful properties observed in artificial as well as homogeneous periodic structures composed of macroscopic elements. Metamaterials possessing a negative index of refraction with single-negative (SNG) or double-negative (DNG) characteristics have spurred a broad interest among researchers with their applications, ranging from the microwave to terahertz regime [1–3]. Antennas are an essential part of wireless communication systems used to transmit/receive signals into space. Although simple by structure, antenna design for microwave and terahertz applications in planar configurations face various challenges such as narrow bandwidth, surface wave loss, poor gain, lower efficiency and less directivity [4–6]. Such problems can be overcome by integrating SNG or DNG metamaterial arrays with antennas leading to various advantages, such as miniaturization of the antenna dimension, enhancement in broadside radiation with increase in gain, and suppression of surface waves [7–9]. Besides antennas, metamaterials are also implemented in periodic structures such as frequency selective surfaces (FSSs) leading to sub-wavelength unit cell dimensions (periodicity $\ll$ wavelength). Metamaterial-based FSSs with a bandstop or bandpass response have numerous applications in the microwave and THz regime [10–12], such as sub-reflectors for antennas, radome, absorber design, polarization reconfiguration, cloaking, and performance enhancement of antennas as a substrate or superstrate.

3.1 Metamaterial-based periodic structures

Metamaterials are structured as periodic arrangements consisting of unit cells that are engineered in a way so that the unit cell dimension, also known as periodicity of

doi:10.1088/978-0-7503-5422-6ch3

the unit cell 'p', is significantly smaller than λ_g, the guided wavelength of the microwave ($p \ll \lambda_g$), as shown in figure 3.1.

This is accomplished so as to average the fields across the unit cells and manipulate the structure as a homogenous medium [13]. Metamaterials, being engineered structures, exhibit negative values of permittivity and/or permeability, which in turn results in various unusual properties such as left-handed wave propagation, backward propagation of waves, electromagnetic cloaking, etc [14, 15].

The epsilon-negative property can be realized using periodic array of thin metallic wires below plasma frequency [16] whereas mu-negative response can be implemented by using an array of split ring resonators (SRR) [17] as shown in figure 3.2(a) and (b), respectively. A combination of the two as shown in figure 3.2(c) lead to negative values of both the parameters [17] and a fabricated prototype of the same is shown in figure 3.3 as presented in [2].

Unlike metamaterials that exhibit a single negative (SNG) or double negative (DNG) response metamaterial-inspired structures should not necessarily exhibit '$\varepsilon < 0$' or '$\mu < 0$'. Such structures are periodic and contain unit cell dimensions much smaller compared to λ, typically on the order of $\lambda/10$ or less [18]. Periodic structures with sub-wavelength dimension such as electromagnetic bandgap (EBG) structures, frequency selective surfaces (FSSs), artificial magnetic conductor (AMCs), metasurfaces can be regarded as metamaterial-inspired structures [18].

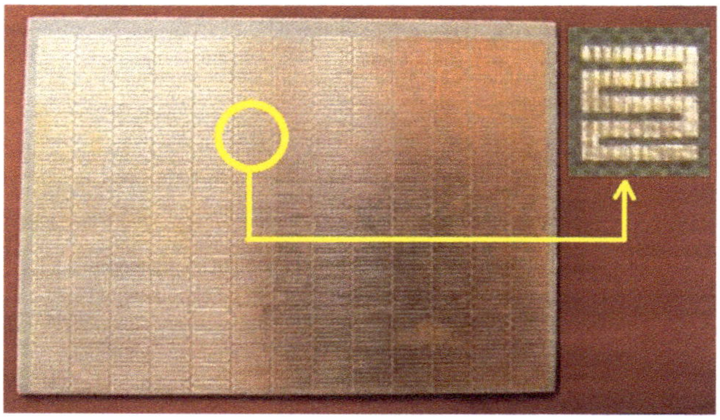

Figure 3.1. Metamaterial-inspired structure with sub-wavelength periodicity.

Figure 3.2. Periodic arrangement of (a) thin conductive wires exhibiting $\varepsilon < 0$, (b) split ring resonator exhibiting $\mu < 0$, and (c) combined thin metallic wires and SRRs for both $\varepsilon < 0$ and $\mu < 0$.

Figure 3.3. Periodic array of thin conductive wires and split ring resonators. Reprinted by permission from Springer Nature Customer Service Centre GmbH: [Springer Nature Link] [MRS Bulletin] [2], Copyright (2011).

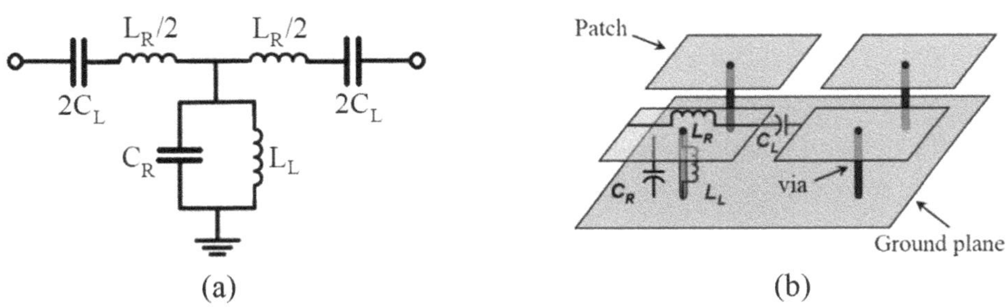

Figure 3.4. (a) Equivalent circuit for symmetrical CRLH unit cell. [19] John Wiley & Sons. [© 2019 Wiley Periodicals, Inc.]. (b) Mushroom type CRLH unit cell. Reproduced from [30]. CC BY 3.0. Reprinted by permission from Springer Nature Customer Service Centre GmbH: Springer Nature, [30], Copyright (2020).

The left-handed structure designed using metallic wires and split rings showing metamaterial properties incurs loss and narrow bandwidth. A modified artificial left-handed structure termed as composite right-/left-handed transmission line (CRLH TL) [19] can overcome these challenges. This is accomplished by incorporating the right-handed (RH) effect into a left-handed (LH) configuration leading to a metamaterial *TL* structure. The equivalent circuit of a symmetrical CRLH *TL* unit cell is shown in figure 3.4(a). The series capacitor C_L and the shunt inductor L_L correspond to the left-handed property while the series inductor L_R and the shunt capacitor C_R represent its right-handed counterpart. The mushroom shaped unit cell [19] as shown in figure 3.4(b) belongs to this type of CRLH metamaterial-based periodic structure.

3.2 MTM-based microwave and THz antenna

Metamaterial-based antenna refers to the class of antennas integrated with different MTM structures such as CRLH *TL*, split ring resonator (SRR) and complementary SRR (CSRR), metasurface, etc., for enhancing various antenna parameters [21, 22].

MTMs are implemented in planar antennas operating in the microwave regime, such as microstrip patch antennas, array antennas, CPW-fed antennas, and electrically small antennas to overcome various design challenges such as narrow bandwidth, poor gain and directivity, smaller radiation efficiency, etc [4–6]. However with the enormous growth in current generation wireless communication enhanced carrier frequencies, higher data rates (100 Gbps) and increased channel capacity became more important, leading to more use of the terahertz (THz) spectrum. Terahertz radiation can augment the performance and decrease the power consumption of cell towers [23]. Moreover, certain materials that do not pass visible and infrared frequency waves become transparent for THz frequency signals. The THz spectrum is more suitable for imaging applications compared to microwave imaging because of good spatial resolution [24]. These lead to an increase in the usage of THz antennas in present communication systems. However, because THz waves are shorter in terms of wavelengths, they suffer from path loss and atmospheric attenuation. This can be overcome with the use of enhanced gain and directivity of the THz antennas, which can be achieved with the implementation of MTM structures in the antennas for THz applications.

3.2.1 MTM-based antennas for microwave applications

MTM-based structures of different configurations as mentioned in section 3.1 are integrated with planar as well as non-planar antennas for a wide range of microwave applications. Such applications include antennas for Wi-Fi, WiMAX, wireless body area networks (WBAN) in the ISM band, zeroth-order resonance antennas, electrically small antennas, microwave imaging, and biomedical engineering [25–27]. MTM structures are chosen based on the applications. For example, CRLH structures are preferred for small antennas and zeroth-order antennas, whereas SRR or CSRR structures are mostly used in antennas for ISM band applications as well as biomedical applications [28, 29]. A detailed discussion on this is presented in this section.

3.2.1.1 MTM antennas for ISM band applications

The ISM band covers the lower end of the microwave spectrum and thus antenna dimensions become large in this band. SRR-based MTM structures are used for miniaturization of such antennas including linearly polarized as well as circularly polarized antennas. Geetharamani and Aathmanesan proposed a simple microstrip patch antenna with the inclusion of complementary split ring resonators (CSRRs) on the patch to resonate at 2.4 GHz for Wi-Fi applications [30]. Two modified split ring resonators (SRRs) are loaded on the ground plane inside a slot, as shown in figure 3.5(a). SRR is used to realize a high-quality factor whereas CSRR is used to improve the antenna return loss and isolation between elements in case of its use as an array antenna. ENG ($\varepsilon < 0$) is realized by feeding the E field normal to the surface of the ring.

It was observed that the MTM-based antenna resonates exactly at 2.4 GHz with a good reflection coefficient of -46.58 and a -10 dB bandwidth of 574 MHz.

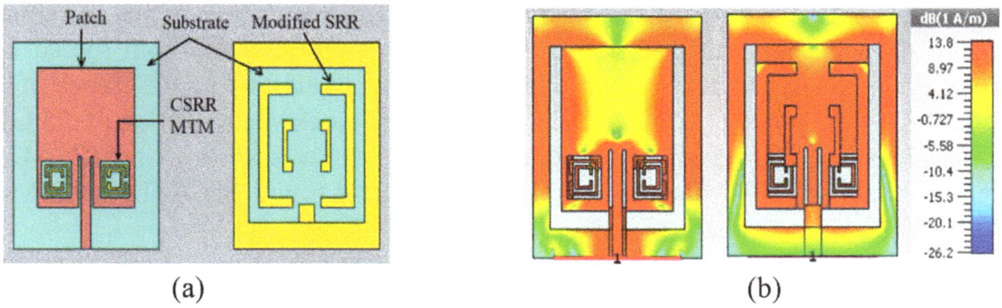

Figure 3.5. (a) Antenna with CSRR MTM (front and back view) and (b) surface current distribution on the antenna planes at 2.4 GHz [30].

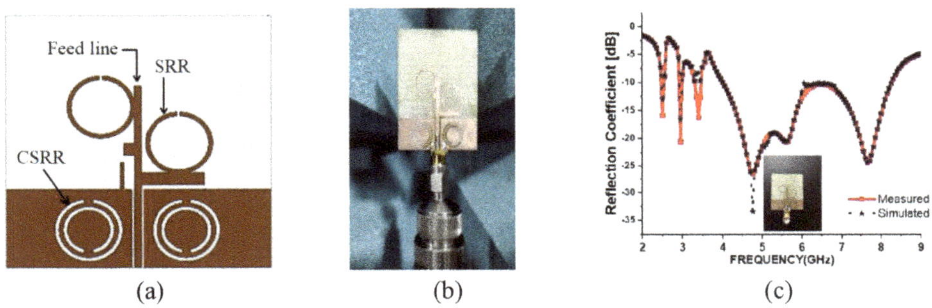

Figure 3.6. (a) Antenna with CSRR MTM (front and back view), (b) surface current distribution on the antenna planes, and (c) fabricated prototype of the antenna. Reprinted from [31], Copyright 2018, with permission from Elsevier.

However, the antenna gain was not enhanced significantly and shown a value of 3.38 dBi. As can be seen from the surface current on the patch and ground plane in figure 3.5(b), the current is mostly excited surrounding the edges and the CSRRs. It was also shown that the current intensity could be improved by modifying the feeding network with the use of a RC network on the feed line.

The MTM-loaded antenna discussed above exhibits linear polarization, whereas for biomedical applications where the person changes body posture frequently as well as in satellite communication applications, circularly polarized antennas are preferred. Venkateswara Rao *et al* have shown that with the use of SRR and CSRR-based MTMs in a CPW-fed monopole antenna, circular polarization can be achieved over quad frequency bands for WLAN, Bluetooth, WiMAX and satellite communication applications [31]. The MTM-loaded antenna as shown in figure 3.6 (a) has CSRR loading on the ground and SRR attached to the feed line along with stub.

The CSRR and SRR structures enhance the antenna impedance bandwidth in the operating frequencies of 2.4 and 5.8 GHz for Bluetooth/Wi-Fi/WLAN, 3.35 GHz for WiMAX, and 7.25–7.75 GHz for X-band downlink satellite communication applications as can be observed in figure 3.6(c). Circular polarization with the axial ratio below 3 dB is achieved due to the asymmetric orientation of the horizontal stubs to the feed line and insertion of SRR. This is also evident from the surface current distribution on the antenna at different phase angles with a difference of 90° in figure 3.7. The current vector rotation in a counter-clockwise direction at 2.5 GHz ensures right-handed circular polarization (RHCP). It was shown that the MTM antenna exhibits RHCP also at 3 GHz, whereas at 4.6 and 6.4 GHz LHCP was observed for the antenna.

3.2.1.2 MTM-loaded small antennas

The antennas with largest dimension equal to or less than one-tenth of a wavelength are referred to as electrically small antennas [32]. Generally CRLH transmission line-based metamaterials, MTMs with epsilon-negative (ENG), mu-negative (MNG), or high-mu shells are used to realize small antennas [26, 33]. Yang *et al* proposed a reconfigurable small antenna based on CRLH transmission line-based MTMs [34] as shown in figure 3.8(a). The antenna has a radiating patch (RP) fed by asymmetric coplanar strips through coupling via a matching patch (MP).

The arrangement of *MP*, *RP* and the ground plane represents a CRLH transmission line. The line between *RP* and ground exhibits multiband operation and a

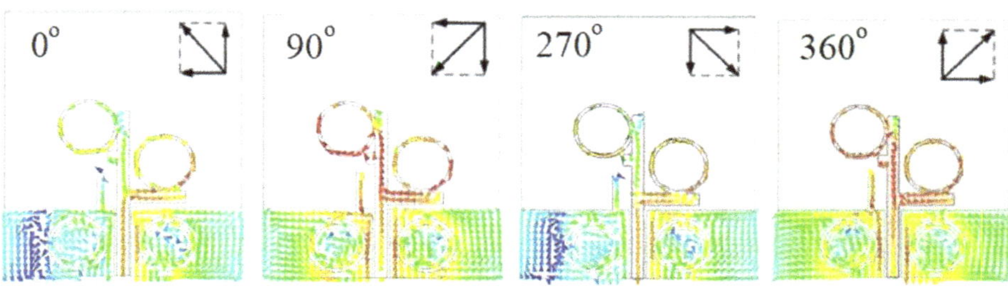

Figure 3.7. Surface current distribution on the MTM antenna at 2.5 GHz at different phases. Reprinted from [31], Copyright 2018, with permission from Elsevier.

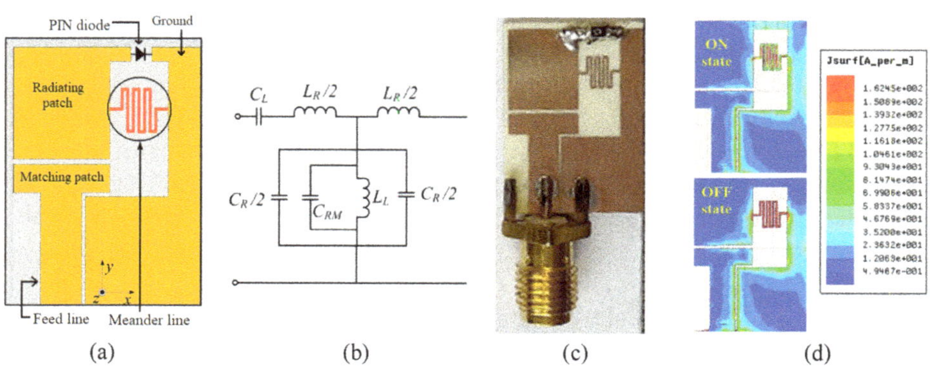

(a) (b) (c) (d)

Figure 3.8. (a) CRLH TL loaded antenna with PIN diode, (b) equivalent circuit for CRLH TL, (c) fabricated prototype of the antenna, and (d) surface current distribution on the antenna for 'ON' and 'OFF' states. Reproduced from [34]. CC BY 4.0.

PIN diode acts as a termination to the CRLH TL. It is shown that by controlling the PIN diode switch (ON and OFF) the operating frequency band can be shifted between zeroth-order resonances, leading to a reconfigurable antenna. An equivalent circuit model for the CRLH TL is proposed, as shown in figure 3.8(b), where *RP* is modeled using L_R, the gap between *MP* and *RP* is modeled using C_L whereas C_R models the space between *RP* and ground plane. The inclusion of the meander line completes the circuit with the shunt components L_L and C_{RM} and the circuit is terminated with the PIN diode equivalent RLC components for both ON and OFF states. The antenna operates in single-band mode (2.08–2.54 GHz) with a peak gain of −2.03 dBi for switch ON state and dual-band mode (1.65–1.84 GHz and 5.35–6.40 GHz) with gain values of −3.12 and 4.24 dBi for switch OFF state. MTM leads to a miniaturized antenna dimension of $0.08\lambda_0 \times 0.11\lambda_0 \times 0.004\lambda_0$ with applications in UMTS, ISM, WIMAX and WLAN frequency bands.

As observed previously, MTM-based electrically small antennas are suitable for applications in the lower end of the microwave spectrum due to low profile, however with the drawback of negative gain. This problem was resolved with the use of CRLH TL and third-iterative fractals (TIF) in designing small antennas, as proposed by Ameen & Chaudhary [26]. Apart from gain enhancement, the MTM-based antenna also exhibits circular polarization (CP). The fabricated antenna with CRLH TL on the front and TIF above partial ground plane on the back side is shown in figure 3.9(a). The adapted 'U'-shaped capacitive slot is etched out from the chamfered semicircular patch in order to realize the series arm of via-less CRLH TL, as depicted in the equivalent circuit (figure 3.9(b)). The curved meander line, rectangular virtual ground plane and the vertical gap between this virtual ground and partial ground account for the CRLH TL shunt components in the circuit. Zeroth-order resonance for this antenna can be controlled by varying the CRLH TL shunt elements as evident from the ZOR frequency [26]:

$$f_{ZOR} = \frac{1}{2\pi\sqrt{L_L \times ((C_R C_V)/(C_R + C_V))}} \tag{3.1}$$

It was shown that the antenna with the inclusion of MTM acquires significant miniaturization of 68.5%, leading to a dimension of $0.11\lambda_0 \times 0.14\lambda_0 \times 0.006\ \lambda_0$ at

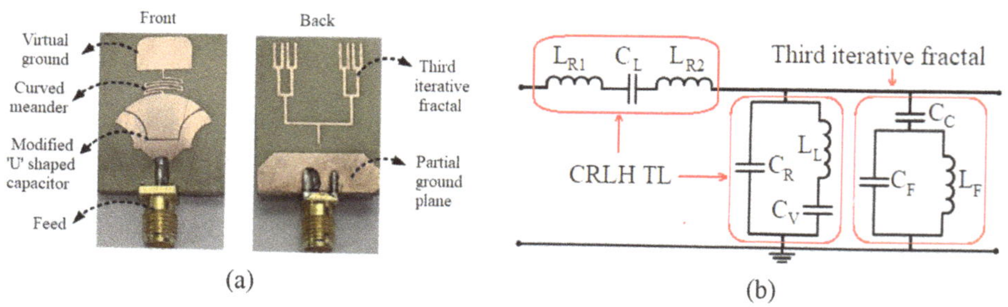

Figure 3.9. (a) Fabricated antenna with CRLH TL on front and third-iterative fractal (TIF) on back and (b) equivalent circuit model for the CRLH TL loaded antenna without TIF on back side. [26] John Wiley & Sons. © 2019 Wiley Periodicals, Inc.

1.635 GHz along with an improved gain of 1.79 dBi. It was proved that the antenna is electrically small with the calculation of quality factor $Q_{\min}$ and fractional bandwidth $\mathrm{FBW}_{\max}$ followed by maximum obtainable gain G_{dBi} using the following equations [26]:

$$\left.\begin{aligned} Q_{\min} &= (ka)^{-1} + (ka)^{-3} \\ \mathrm{FBW}_{\max} &= \frac{\mathrm{VSWR} - 1}{Q_{\min}\sqrt{\mathrm{VSWR}}} \\ G_{\mathrm{dBi}} &= 10\log_{10}[(ka)^2 + 2(ka)] \end{aligned}\right\} \tag{3.2}$$

where $k = 2\pi/\lambda$ and 'a' stands for the radius of a sphere encircling the largest dimension of the MTM-loaded antenna. The authors proved that with $ka = 0.59$ the proposed antenna is electrically small. The theoretically obtained values were found to be in close agreement with the measured values for these parameters. These characteristics along with the CP induced by TIF make the antenna suitable for L-band satellite communication.

3.2.1.3 MTM antennas for microwave imaging

Microwave imaging has a wide range of applications, including nondestructive testing and evaluation that involves determination of material properties, condition monitoring, medical imaging, concealed weapon detection at security check points, structural health monitoring, and explosive detection using ground penetrating radar (GPR) [35–37]. Besides various conventional methods the use of ultra-wide band antennas has become popular in microwave imaging especially for medical diagnosis and GPR [38]. The imaging is accomplished with the analysis of transmitted and received short duration pulses from UWB antenna because UWB signals exhibit better resolution and large depth of penetration. Metamaterials are incorporated in UWB antennas for enhanced gain and directivity retaining the wideband response [39, 40].

Gopikrishnan *et al* have demonstrated a MTM-based eight-slot Vivaldi antenna for the detection of hidden crevasses in glacier terrain using microwave imaging [41]. The antenna operates over a wide range of 1.5–10.5 GHz, covering the lower as well as higher end of the spectrum leading to enhanced resolution of the image. A significant number of spiral shaped anisotropic zero-index metamaterial unit cells are loaded on the Vivaldi antenna, as shown in figure 3.10(a).

The antenna slots are fed by a 1–8 wideband power divider based on the hebyshev T-junction impedance transformation technique [42] as can be observed the same figure. The MTM is responsible for augmentation of characteristics in lower band of 1.5–6 GHz. It was observed that the MTM-loaded antenna ibits a directivity up to 16 dBi and cross-polarization level reduced up to −20 dB ble for microwave imaging. The proposed experimental setup, including a r network analyzer (VNA) with time-domain facility connected to the MTM a and associated RF components, can be seen from figure 3.10(b). The s have used polystyrene foam of various widths (20, 29, and 40 cm) emulating

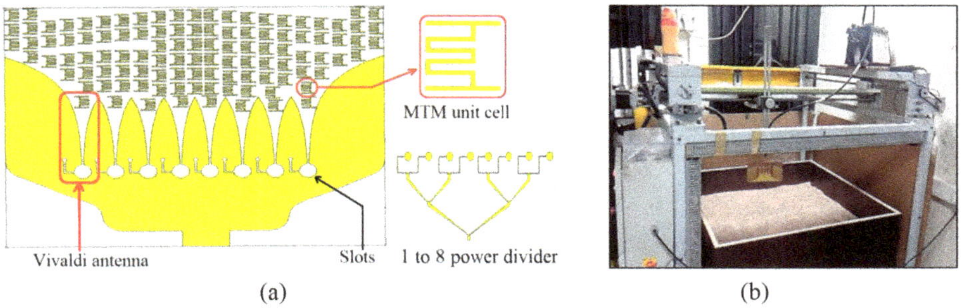

Figure 3.10. (a) Schematic of the eight-slot Vivaldi antenna loaded with MTM (front) and 1–8 power divider (back). (b) Experimental setup for the sample crevasse imaging [41] 2019, reprinted by permission of the publisher (Taylor & Francis Ltd, http://www.tandfonline.com.)

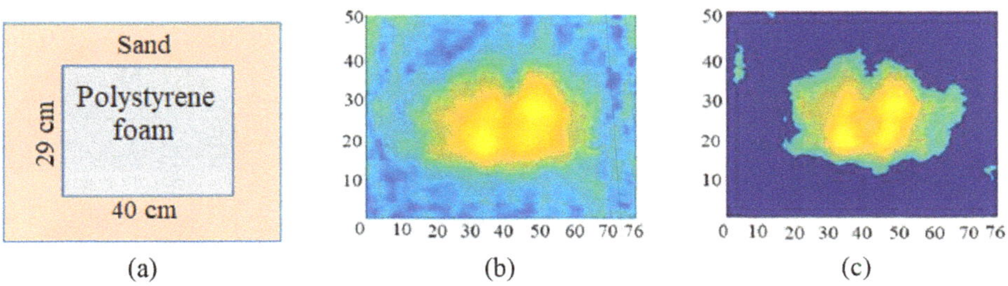

Figure 3.11. (a) Replicated arrangement for crevasse in glacier using sand and foam, (b) raw data, and (c) processed data for reflectivity image of the sample. [41] 2019, reprinted by permission of the publisher (Taylor & Francis Ltd, http://www.tandfonline.com.)

the crevasses and sand for the surroundings, as shown in figure 3.11(a). The contrast in the magnitude of time-domain reflection plot for the sample in the scanning area was observed and it was concluded that the region with polystyrene foam (emulating an air gap) has the highest contrast, as evident from figures 3.11(b) and (c). This high contrast area in the reflectivity image indicates the crevasse location.

Microwave imaging is extensively used in biomedical engineering for medical diagnosis, especially for fatal disease such as tumor detection, breast cancer detection, neuroimaging for brain tumor detection, etc [43–45]. In most cases, array antennas are used for medical imaging where one antenna is used as transmitter and others are used as receiver. A typical imaging system is shown in figure 3.12. The biomedical sensors along with the antenna array are mounted on the body where each antenna is connected to a multiplexer. The multiplexer automatically switch one antenna in transmit mode and the others in receive mode. The signals are fed to the signal processing unit for image analysis, which is then fed to a display unit.

Alibakhshikenari *et al* proposed a MTM-inspired planar antenna array for detecting tumors in biological tissues [45]. In the proposed model 8 element antenna arrays each with an MTM-inspired patch, is placed over a breast model as partially shown in figure 3.13(a). Here square-shaped concentric rings of sub-wavelength

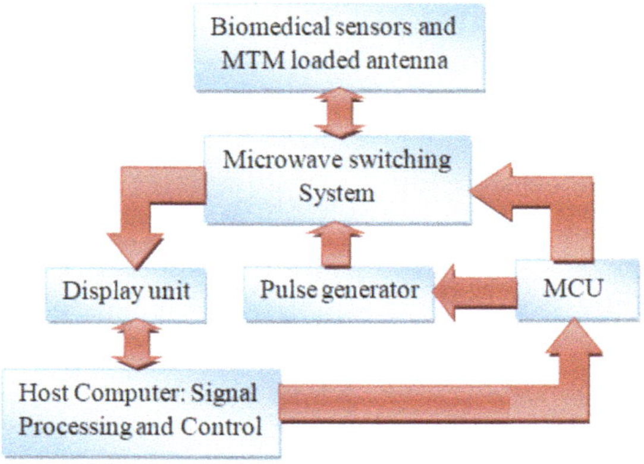

Figure 3.12. Microwave imaging system for tumor detection.

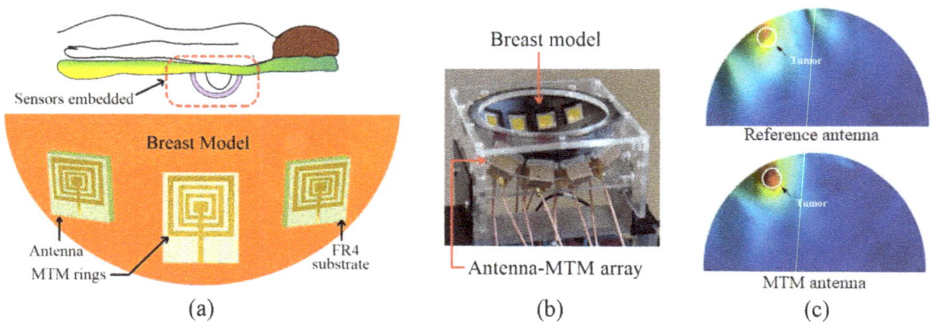

Figure 3.13. (a) Breast model with MTM-inspired antenna array, (b) breast model using hemispherical resin cup and antenna array, and (c) reconstructed image of tumor for antenna array with and without MTM. Reproduced from [45].

dimension are used as MTM on the antenna patch. The array structure operating in 2–12 GHz exhibits an enhancement in antenna gain and efficiency by 4.8 dBi and 18% with the use of MTM leading to an efficient imaging. With the S11 below −20 dB the antenna becomes receptive to even weak signals and minimizes distortion in the short duration pulse-trains transmitted for scanning. The breast tissue model is made using a hemispherical resin cup with the same electric and optical properties for measurement as can be seen from figure 3.13(b) incorporating the antenna array and feeding network. A spherical tumor with the diameter of 5 mm was located near the skin tissue inside the breast model.

The reconstructed images of the fields scattered from a standard patch antenna array and that from the proposed MTM-based antenna array were investigated at 5.5 and 12 GHz as shown in figure 3.13(c). As observed the microwave pulses are mostly reflected from the region of the breast with a higher ε_r than normal breast tissue (ε_r of the tumor is 54 and breast tissue is 10). The tumor can be detected more prominently in the case of a MTM-based antenna array.

3.2.2 MTM-based antennas for THz applications

In the past few years THz antennas have been extensively used for environmental, biomedical, and industrial applications [6, 46–48]. Antennas made of conducting materials function well at microwave frequencies; however, their radiation characteristics degrade radically in the THz frequency or far-infrared region because of the skin effect and thermal losses [6]. These challenges can be overcome with the use of graphene, which is an allotrope of graphite, with carbon atoms bonded in a 2D honeycomb crystal grid. The use of graphene in designing THz antennas leads to miniaturization of the antenna dimension, however smaller dimensions reduces the antenna efficiency by around 5% or more [47]. This can be compensated with the use of negative index ($n < 0$) or low index ($n < 1$) materials, such as metamaterials, by enhancing radiation parameters of the THz antenna including gain, directivity and radiation efficiency [47].

3.2.2.1 MTM-loaded graphene-based THz antenna

Although metallic antennas exhibit desired performance in the near-infrared region, poor internment of metal surface plasmon polariton (SPP) waves at THz frequencies limits their application [46]. Graphene on the other hand supports the propagation of highly-confined SPP waves at transverse magnetic (TM) mode with increased propagation constant compared to that of free-space. This in turn reduces the SPP wavelength, leading to miniaturization of antenna dimensions, but at the same time degrades radiation efficiency of the antenna. Metamaterials possessing unusual properties with a relative permittivity less than that of free space can be useful in increasing the antenna efficiency [47].

Graphene being a perfectly 2D material can be characterized with the help of its surface conductivity using the Kubo formula $\sigma(\omega, \mu_c, T)$; where ω is the radian frequency, μ_c the chemical potential controlled either by applied gate voltage or chemical doping, and T the ambient temperature (300 K). Graphene's conductivity is evaluated as $\sigma = \sigma_{\text{intra}} + \sigma_{\text{inter}}$, among which σ_{inter} is not significant at the lower end of the THz spectrum and the conductivity mostly depends on σ_{intra}, given by the following equation [46, 49]:

$$\sigma_{\text{intra}} = \frac{2e^2 k_B T}{\pi \hbar^2} \frac{j}{\omega + j\tau^{-1}} \left[\ln \left\{ 2 \cosh \left(\frac{\mu_c}{k_B T} \right) \right\} \right] \tag{3.3}$$

Here e is the charge of electron (1.6×10^{-19} C), k_B is Boltzmann's constant (1.38×10^{-23} J K^{-1}), $\hbar$ is the reduced Planck constant, τ is the relaxation time (1 ps). The propagation properties of graphene depend on its conductivity. The complex propagation constant of the surface wave k_ρ in case of an infinite graphene layer located between two dielectrics characterized by ε_{r1} and ε_{r2}, can be estimated using the following equation given by [47]:

$$\frac{\varepsilon_{r1}}{\sqrt{(k_\rho/k_0)^2 - \varepsilon_{r1}}} + \frac{\varepsilon_{r2}}{\sqrt{(k_\rho/k_0)^2 - \varepsilon_{r2}}} = j\sigma\eta 0 \tag{3.4}$$

Here k_0 and η_0 are the wave number and wave impedance in vacuum, while σ is graphene surface conductivity (1). The wavelength λ_{SPP} of the SPP wave in TM mode can be evaluated [47] from the complex propagation constant k_ρ by using the relation $\lambda_{SPP} = 2\pi/\text{Re}[k_\rho]$. It is apparent from the above discussion that MTM-loaded graphene-based antennas are potential candidates for THz applications. Amanatiadis *et al* proposed a graphene-based patch antenna for THz applications at 1.67 THz [47]. Silica (SiO_2) with $\varepsilon_r = 3.3$ is used as the substrate and the L/W ratio is kept at 2 with the length (L) chosen optimally as 24 μm for maximum efficiency of 21.2%. However, it was also observed that with an epsilon-near-zero (ENZ) medium as the substrate and $L = 48.7$ μm the antenna exhibits an increased efficiency of 36%. An ELC resonator-based metamaterial was incorporated in the antenna to realize both ENG (ε negative) and ENZ behaviour. An array of ELC resonators is used beneath the graphene sheet in the antenna to implement the MTM property as the field strength in the substrate is larger near the graphene sheet due to SPP propagation. The MTM array made of ELC resonator, as shown in figure 3.14 (a), is fabricated within the antenna substrate with several SiO_2-based vertically placed slabs, as depicted in figure 3.14(b).

In order to accommodate more MTM cells, the ELC resonator is modified to an OE1 resonator incorporated with an interdigital capacitor and inductive spirals, as can be observed from figure 3.14(a), leading to miniaturization of the unit cell dimension. The electric field strength in the plane orthogonal to MTM cells is more intense in the centre where graphene and MTM cells are accommodated together. The use of MTMs in the graphene-based antenna leads to a 4× increase in the efficiency (3.9%–16.6%), making it suitable for THz applications.

Graphene can be used to design reconfigurable THz antennas due to its tunable surface conductivity as indicated in Kubo's formula [46, 49], given in equation (3.3). The conductivity is generally tuned by changing the chemical potential (μ) of graphene that can again be adjusted by applying a *DC* bias on the graphene.

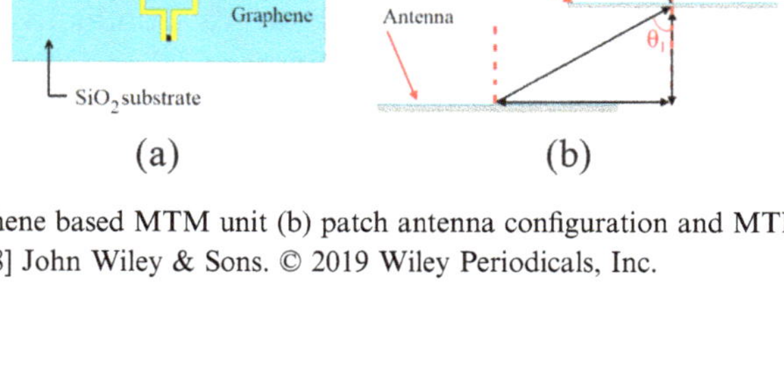

Figure 3.14. (a) Graphene based MTM unit (b) patch antenna configuration and MTM layers as superstrate above the antenna. [48] John Wiley & Sons. © 2019 Wiley Periodicals, Inc.

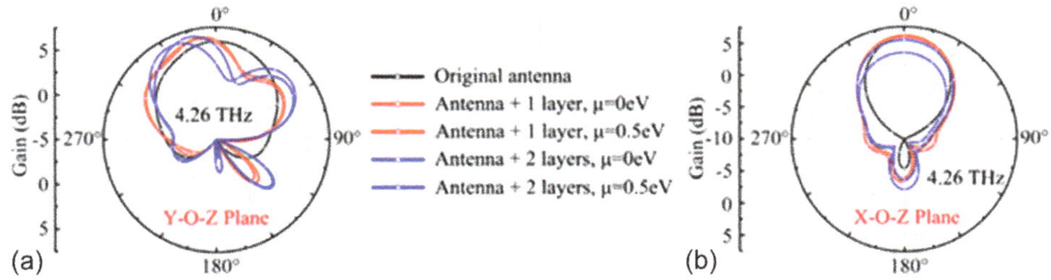

Figure 3.15. Radiation plot of antenna-MTM for the variation of chemical potential under different situations in the (a) YOZ plane and (b) XOZ plane [48] John Wiley & Sons. © 2019 Wiley Periodicals, Inc.

Luo *et al* proposed a graphene-based tunable MTM unit, as shown in figure 3.15(a), to realize a negative refractive index [48]. The MTM consists of a metallic resonator composed of a modified square loop surrounded by four square loops.

Graphene is embedded in the four loops, as can be seen from the schematic. The MTM-based on a SiO_2 substrate exhibits a negative refractive index at 4.28 THz for a chemical potential (μ) of 0.5 eV, however the value is more negative for 0 eV at 4.26 THz. The desired value of μ can be reached by applying a DC bias between the MTM and p-type Si placed below SiO_2. It was shown that the refractive index (n) can be estimated from reflection (S_{11}) and transmission (S_{21}) coefficients of the MTM obtained using an EM simulator using the following relation (3.5) [48]:

$$n = \frac{1}{kd}\arccos\left[\frac{1}{2S_{21}}\left(1 - S_{11}^2 + S_{21}^2\right)\right] \tag{3.5}$$

Here, k is the free space wave number and d is the effective thickness of the MTM. A negative value for both the permittivity and permeability is realized with the MTM leading to the left-handed property. The MTM is embedded in a simple patch antenna as superstrate with the arrangement shown in figure 3.15(b). The incident and refraction angles are on the same z-axis because of $n_{air} > 0$ and $n_{MTM} < 0$. This graphene-based MTM loading enables dynamic beam tilting of the antenna, making it suitable for secured and high-speed wireless communication. The antenna beam exhibits a peak gain of 5.9 dB at 0° without MTM and it was observed that for the antenna with MTM loading the beam gets tilted with a variation in chemical potential μ. As depicted in figure 3.15(c), an antenna with single-layer MTM tilts the beam by 11° for $\mu = 0.5$, whereas for the same μ and dual-layer MTM the beam gets tilted by 27°. Dynamic beam tilting of THz antennas is also applicable in compensating high path loss faced by THz signals due to shorter wavelength [48].

3.2.2.2 MTM-based THz antenna for biomedical applications

In recent years, MTM-based THz antennas have been used extensively for biomedical applications, such as medical imaging for the detection of blood clots, brain tumors, and oral, skin, and breast cancer [50–52]. Metamaterial structures

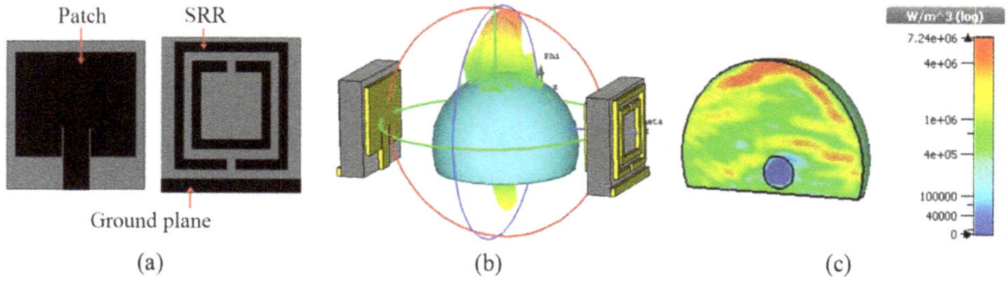

Figure 3.16. (a) Antenna patch and SRR-based MTM above ground plane, (b) experimental setup for cancer detection using two antennas, and (c) power plot for the breast model with tumor. Reprinted by permission from Springer Nature Customer Service Centre GmbH: [Nature] [SN Applied Sciences] [52], Copyright (2019).

such as the split ring resonator (SRR), ELC resonator and graphene-based hyperbolic MTMs are used for this purpose. Geetharamani and Aathmanesan proposed an SRR-based patch antenna operating in the THz regime for breast cancer detection [52]. A ground plane made of gold with the SRR above it is incorporated on the back side of the patch antenna, as shown in figure 3.16(a). The antenna resonates at 1 THz with the reflection coefficient of −35 dB and −10 dB bandwidth of 0.37 THz.

The CSRR acts as a resonant magnetic dipole, which is excited by an axial electric field to realize the MTM property, which is negative permittivity for the proposed design [53]. With the MTM the antenna shows a peak gain of 20 dBi, making it suitable for detection of weak reflected signals from human body tissue. Breast cancer detection works by evaluating the difference between a normal tissue and a malignant tissue by using the variation in dielectric properties of the normal and malignant tissue [54]. The malignant tissue is identified by analyzing the impulse response of two MTM-based antennas placed on both sides of the sample, as shown in figure 3.16(b). A relative permittivity of 2.41 and 3.18 are chosen for the breast model and tumor, respectively. With the investigation of the power pattern it was observed that the power is distributed uniformly over the breast tissue, whereas in the presence of a tumor the distribution is not even and some portion of the radiated power gets absorbed by the tumor, leading to deflections in power distribution. Moreover, the tumor detection was confirmed from the time-domain analysis of the THz signal that shows the occurrence of an impulse in the amplitude response.

Poorgholam–Khanjari and Zarrabi have proposed [54] the use of graphene as a hyperbolic metamaterial in a wideband Vivaldi antenna operating between 0.5 and 2 THz for cancer detection. The antenna loaded with parasitic elements and MTM is shown in figure 3.17(a).

Quartz substrate with $\varepsilon_r = 3.75$ and thickness of 10 μm is used as the substrate for the antenna. The antenna is integrated with hyperbolic metamaterial for the purpose of enhancing its performance. Hyperbolic metamaterials have indefinite dispersion and the sign of either permittivity or permeability component is opposite to the other one. It was shown through equation (3.6) for obtaining ε_r of graphene that graphene can be used as a hyperbolic MTM.

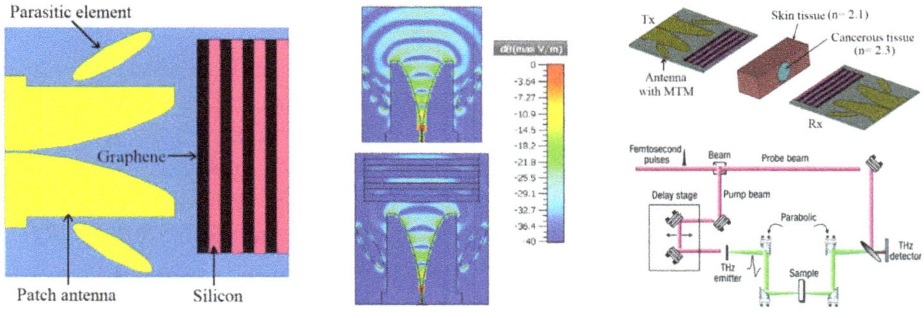

Figure 3.17. (a) Graphene-based hyperbolic metamaterial loaded Vivaldi antenna, (b) electric field on the antenna with and without graphene-based MTM, and (c) cancer detection using TDS system. Reprinted from [54], Copyright (2021), with permission from Elsevier.

$$\varepsilon_r = 1 - j\frac{\sigma}{\omega\varepsilon_0 d} \tag{3.6}$$

A combination of graphene as an MTM with silicon as a dielectric is incorporated in front of the antenna patch, among which graphene enhances the antenna bandwidth in addition to reducing gain, whereas the dielectric enhances the antenna gain. This can be observed from figure 3.17(b) showing the electric field distribution on the antenna with and without hyperbolic MTM-dielectric. A peak gain up to 13 dBi is obtained for the potential of 0.5 eV applied to graphene. For a chemical potential of 0.7 eV the graphene-based antenna exhibits return loss below −10 dB; however, with the gain of 11 dBi. The antenna is used to detect cancer cells in the human body as per the arrangement shown in figure 3.17(c) with two antennas labelled as transmitter (Tx) and receiver (Rx) with the cancer and skin tissue placed in between. The tumor detection is done based on the difference between transmission coefficients for the tissue with and without tumor cells.

3.3 MTM-based FSSs for microwave and THz applications

Frequency selective surfaces were once being specifically used as dichroic spatial filters and sub-reflectors, but have acquired attention of the industry for a wide variety of microwave and THz applications nowadays. However, with the ever-growing demand of miniaturization of the equipment dimensions with the use of the upper end of the microwave spectrum and mm-waves, FSS structures based on metamaterials have taken an important role. MTM-based FSS structures designed at microwave, mm-wave and THz frequencies are extensively used in realizing bandpass or bandstop filters [55, 56], absorbers [57, 58], polarizers [59, 60], and radomes [61], as well as for the enhancement of antenna radiation parameters including gain, directivity, etc., by integrating the MTM-based FSS as superstrate [62, 63]. In this section, various applications of MTM-based FSS structures in the microwave and THz regimes are discussed.

3.3.1 MTM-based FSS structures for filtering applications

The realization of metamaterial-inspired FSSs with sub-wavelength unit cell dimensions was initially proposed by Behdad and his research group in the microwave spectrum [64, 65]. Most of these FSS structures are composed of an array of metallic patches cascaded with the array of square-shaped apertures suitable for application as wireless filters with bandpass response. The filters are often integrated with microwave antennas to realize filtenna [66]. SRR, CSRR and meander line MTM structures are used for realizing microwave absorbers [57]. Conventional FSS unit cells loaded with interdigital capacitors as well as CRLH transmission line-based FSS structures are generally chosen for conformal applications [67].

MTM-inspired FSS structures can be realized as both bandpass and bandstop filters in the microwave regime. Higher-order bandpass responses can be achieved by using multilayered FSS structures, the unit cell of which consists of non-resonant metallic patches on the top and bottom exhibiting capacitance, and a metallic grid exhibiting inductance in the middle [65]. Second-order bandpass responses can be realized with this three-layered configuration, whereas higher-order responses can be achieved by cascading a larger number of FSS layers. Al-Joumayly and Behdad proposed a method of designing such MTM-inspired FSS structures using an equivalent circuit method [64]. The dimensions of the non-resonant unit cell can be estimated from the values of LC components in the circuit and certain desired parameters such as resonating frequency, bandwidth, coupling coefficient, etc. The proposed cascaded FSS along with the MTM-inspired unit cells are shown in figure 3.18(a). A simplified equivalent circuit model is shown in figure 3.18(b) (on top) where C_1 and C_3 model the patch type layers and L_2 models the grid layer. Dielectric substrates of height $h_U = h_L = h$ between the layers are modeled using transmission lines composed of series L (L_{S1} and L_{S2}) and shunt C (C_{S1} and C_{S2}) proposed by Telegrapher [86]. The free space on both ends of the FSS are represented by the product of free space impedance (Z_0) and normalized load impedance (r_1 and r_2).

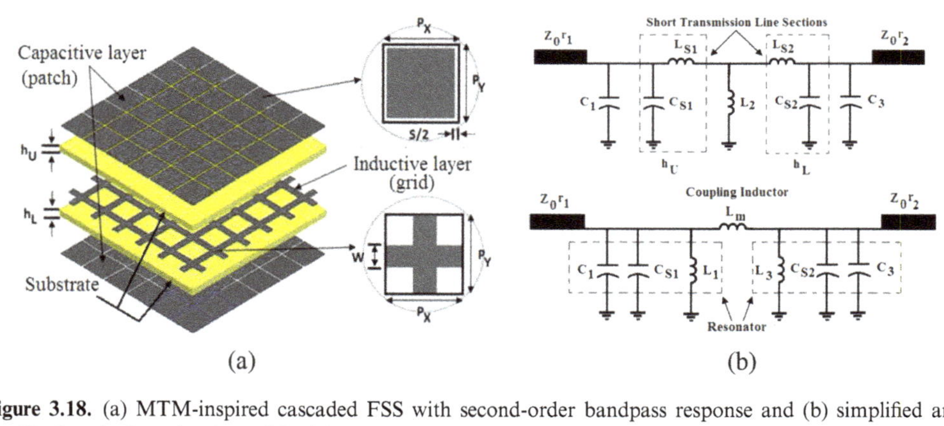

Figure 3.18. (a) MTM-inspired cascaded FSS with second-order bandpass response and (b) simplified and modified equivalent circuit model of the FSS. Reproduced from [63]. CC BY 4.0.

Table 3.1. Parameters for the desired filter response [63].

	Normalized loaded quality factors			Normalized impedances	
Filter type	q_1	q_2	Normalized coupling coefficient k_{12}	Source (r_1)	Load (r_2)
Butterworth	1.4142	1.4142	0.707 11	1	1

In the design process, initially these LC components are evaluated using the desired resonating frequency f_0 (GHz), fractional bandwidth δ (bandwidth/f_0) and certain other parameters for the desired filter response (e.g., the parameters for the Butterworth filter response of second order are given in table 3.1). The circuit is thereafter completed with the estimated values of L_2, C_1 and C_2 obtained from equations (3.7)–(3.9) as well as L_{S1}, L_{S2} ($=\mu_0\mu_r h$) and C_{S1}, C_{S2} ($=\varepsilon_0\varepsilon_r h$) [64, 65].

$$L2 = \frac{Z_0}{(2\pi f_0)k_{12}} \times \frac{(k_{12}\delta)^2}{1-(k_{12}\delta)^2} \times \sqrt{\frac{r_1 r_2}{q_1 q_2}} \quad (3.7)$$

$$C1 = \frac{q_1}{(2\pi f_0)Z_0 r_1 \delta} - \frac{\varepsilon_0\varepsilon_{r1}h_U}{2} \quad (3.8)$$

$$C3 = \frac{q_2}{(2\pi f_0)Z_0 r_2 \delta} - \frac{\varepsilon_0\varepsilon_{r2}h_L}{2} \quad (3.9)$$

In the next and final step the dimensions of the MTM FSS unit cells are calculated from the desired LC values of the equivalent circuit by using equations (3.10) and (3.11) [64]. It is considered that from a practical point of view, spacing between the consecutive patches 's' is to be chosen more than or equal to 0.3 mm below which the fabrication becomes inaccurate.

$$C1 = C3 = C = \varepsilon_0\varepsilon_{\text{eff}}\frac{2P}{\pi}\ln\left(\frac{1}{\sin\frac{\pi s}{2P}}\right) \quad (3.10)$$

$$L2 = L = \mu_0\mu_{\text{eff}}\frac{P}{2\pi}\ln\left(\frac{1}{\sin\frac{\pi w}{2\text{P}}}\right) \quad (3.11)$$

It is observed during the design process that after obtaining the periodicity of the unit cells 'P' and grid width 'w' a fine tuning is necessary to obtain the desired response. The unit cell periodicity is found to be of sub-wavelength dimension on the order of $0.15\lambda_0$, which is a distinct advantage of metamaterials. The equivalent circuit can be justified to model a second-order bandpass filter with transformation to the modified circuit shown in figure 3.18(b) (bottom). With the 'T' network composed of L_{S1}, L_{S2} and L_2 converted to a π network composed of L_1, L_m, and L_3

it is observed that the circuit has two LC resonators leading to a second-order bandpass filter with two transmission poles. This can be confirmed from the transmission response of the FSS obtained through an EM simulator, as shown in figure 3.19. It is evident from the study of angular variation of the incident plane wave that the MTM-inspired FSS exhibits a stable response.

It was later studied by Kou *et al* that [65] the MTM-inspired patch-grid-patch FSS so far discussed contains spurious transmission windows or harmonics outside the operating band (as shown in figure 3.20(b), red lines) that are undesired. They have proposed a harmonic-suppressed frequency selective metasurface element (HS FSM) using square loop elements with the periodicity of around 0.11λ at bandpass operating frequency.

They designed the unit cell by cascading two such FSM layers, each composed of 2×2 square loop elements, as shown in figure 3.20(a). The individual layers offering a bandstop response are separated by an air gap to realize large stop band in the frequency range overlapping with the harmonics (beyond 16 GHz). The FSM offers -20 dB harmonic suppression in the range of 16.8–26.8 GHz. The transmission band at 10 GHz was retained after incorporation of the HS FSM and harmonics

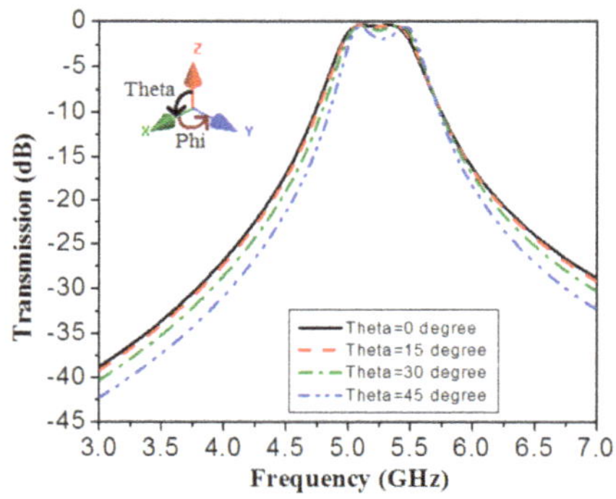

Figure 3.19. Second-order bandpass response of the multilayered MTM-inspired FSS. Reproduced from [63]. CC BY 4.0.

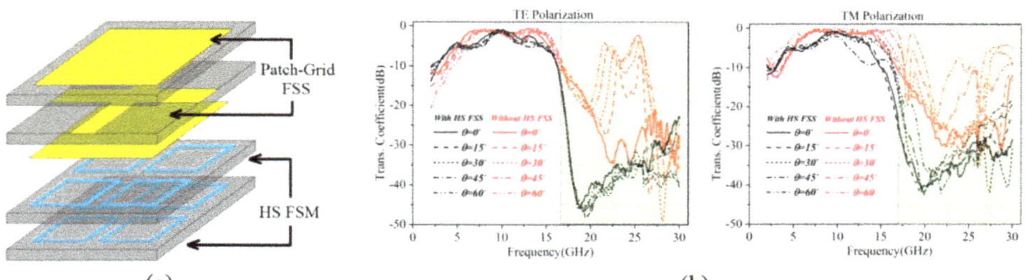

Figure 3.20. (a) Integration of the harmonic-suppressed frequency selective metasurface with conventional patch-grid FSS. (b) Transmission resonance of the FSS with and without FSM in TE and TM planes. Reproduced from [65]. CC BY 4.0.

beyond 16 GHz were suppressed even by up to 40 dB for various angles of incidence of the plane wave for both TE and TM polarization, as can be observed in figure 3.20(b).

Metamaterial-based FSS structures can be used as bandpass filters in the THz regime using the method discussed above [68] or with the use of graphene in addition to conducting materials [56]. Graphene has the advantage of frequency tuning by varying its conductivity with the variation in chemical potential, as discussed in section 3.2.2. MTM-based THz filters have applications such as THz sensing because of properties like miniature size, low cost, and ultrathin thickness [69]. Metallic structures are realized using SiO_2, Si_3N_4, Al, etc. Sorathiya *et al* proposed a MTM-inspired graphene-based FSS in the far-infrared or THz region (GIFSS) with tunable resonance [56]. They have realized the normal (N) and complementary (C) version of the unit cell followed by rotation of the unit cell by +30° and −30°, as shown in figures 3.21(a) and (b). As shown, the unit cell consists of a cross shaped graphene sheet based on a silica based substrate ($\varepsilon_r = 2.25$ and $H = 1.5$ μm) with the periodicity of 7600 nm. The transmittance, reflectance, phase variation, and effective refractive index were investigated for the structure in the 1–5 THz range against the variation of Fermi voltage as well as the chemical potential (figures 3.22(a)–(c)). It was observed that the resonance frequency can be varied depending on the Fermi energy, leading to a wideband transmittance response in the range of 2–4 THz with more than 60% of amplitude. The GIFSS exhibits a negative refractive index at the certain resonance peaks that evident the metamaterial property of the structure.

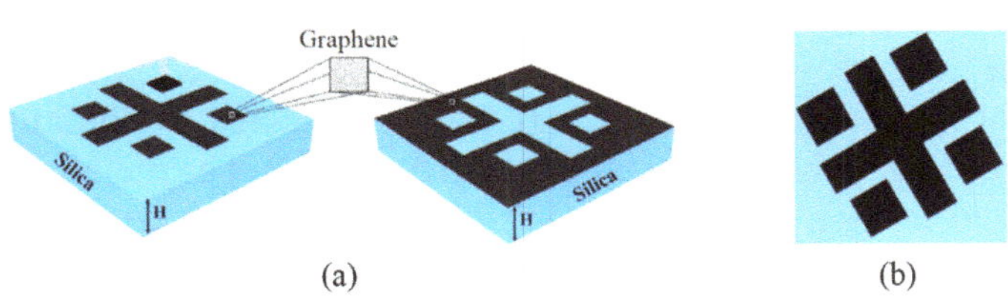

(a) (b)

Figure 3.21. (a) Graphene based MTM-FSS unit cell (N and C) (b) unit cell rotated by −30°. Reprinted by permission from Springer Nature Customer Service Centre GmbH: Springer Nature, [56], Copyright (2022).

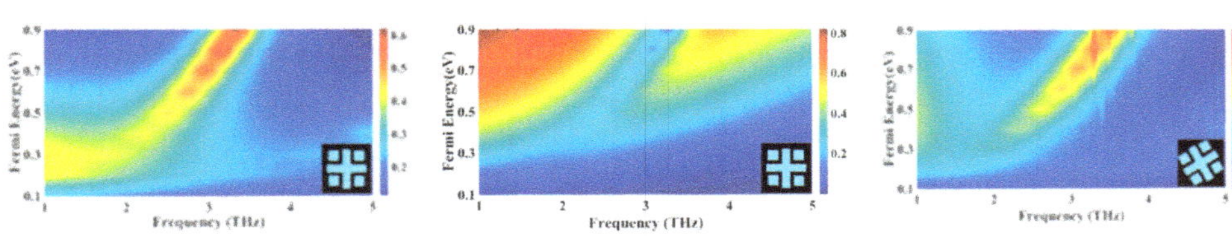

(a) (b) (c)

Figure 3.22. (a) Transmittance at 0° GIFSS-C (b) Reflectance at 0° GIFSS-C (c) Transmittance at −30° GIFSS-C. Reprinted by permission from Springer Nature Customer Service Centre GmbH: Springer Nature, [64], Copyright (2022).

3.3.2 MTM-based FSS structures as absorbers

Traditional microwave absorbers are based on pyramidal absorbers with radiation-absorbent material (RAM) comprising of a rubberized foam material impregnated by mixing carbon and iron [70]. These are frequently used in various applications, including anechoic chambers for measurements of microwave devices and equipment. However, due to larger dimensions they are not suitable for applications such as stealth military aircraft, radomes, etc. Metamaterial-based FSS structures can be used to realize planar microwave absorbers with ultrathin sub-wavelength unit cell dimensions [10, 57, 71] that are even flexible at times suitable for conformal applications [72]. Most of these absorbers are double layered with the lower layer fully conducting.

Mishra *et al* proposed a MTM-inspired FSS-based ultrathin absorber with multiband response at 4.19, 9.34 and 11.48 GHz [57]. The absorber consists of two resonators (FSS A and FSS B) in the unit cell above FR4 substrate ($\varepsilon_r = 4.4$ and loss tangent = 0.02) and 35 μm thick full copper on the back side as shown in figure 3.22(a) with the fabricated prototype in figure 3.22(b). Concentric rings with splits and connected via metallic strips are used to form the unit cell responsible for absorption peaks at resonating frequencies, as depicted in figure 3.22(c). It can be observed from the response that the MTM-based FSS exhibits high absorption of EM waves at resonances where reflectance is nearly zero. High absorptivity results from smaller reflections as well as transmission level, as given by equation (3.12).

$$A = 1 - |S_{11}|^2 - |S_{21}|^2 \tag{3.12}$$

Apparently, due to the presence of ground plane transmission, coefficient S_{21} becomes close to zero and thus with minimum reflection the structure exhibits absorptivity of nearly unity, or 100%. It was investigated by the authors that the absorber exhibits 99% absorptivity at all the frequencies. The metamaterial property was investigated using a dispersion graph obtained through equation (3.13) and it was evident that slope of the graph is negative at all the absorption bands ensuring that all the bands lie in the left-handed region leading to anti-parallel phase and group velocities.

$$\beta_d = \cos^{-1}\left(\frac{1 - S_{11}S_{22} + S_{12}S_{21}}{2S_{21}}\right) \tag{3.13}$$

The absorption nature of the structure can be well understood from the electric field and surface current distribution on the unit cell, as depicted in figure 3.23 at 4.19 GHz. The current direction on the top layer is opposite to the bottom layer, creating a circulating current that is responsible for magnetic excitation, whereas an electric field concentrated on the upper layer leads to electric excitation. The electric and magnetic excitation together lead to maximum absorption. The absorber is polarization independent with respect to linear polarization due to symmetry in the unit cell design. The absorber being MTM-inspired has an extremely low profile unit cell on the order of $0.11\lambda_0 \times 0.11\lambda_0$ with ultrathin thickness of $0.011\lambda_0$. However,

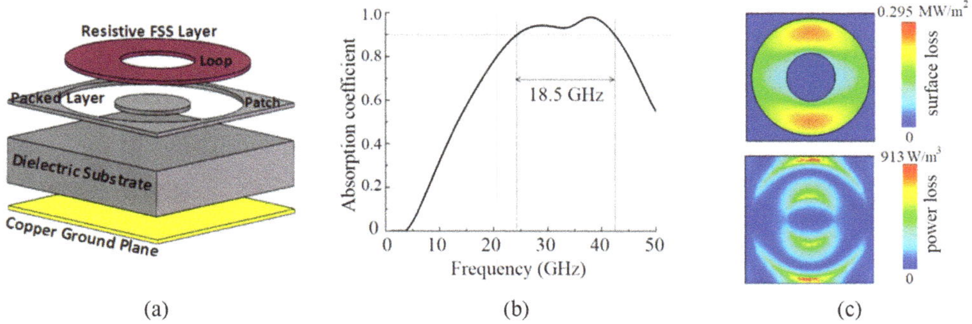

Figure 3.23. (a) Three dimensional view of the MTM based absorber unit cell (b) Absorption coefficient of the FSS based absorber (c) surface loss and power loss on the absorber at 28.9 GHz. Reproduced from [10].

the absorber exhibits narrow bandwidth on the order of 4% in all the operating bands.

Absorbers based on MTMs designed using only metallic FSS layers, as discussed above, generally offer a narrowband response, whereas applications involving aircraft, radomes, etc., need absorption over a wide bandwidth [10, 71]. This can be achieved with the combination of resistive film and metallic patch or with the incorporation of lumped resistors in the unit cell. Such modifications increase the resistance of the resonant structure, leading to a wideband response. Lv *et al* proposed a wideband, ultrathin and polarization independent MTM-inspired absorber with the use of resistive FSS [10]. The unit cell consists of cascaded layers including a ring-shaped resistive FSS embedded into a metallic patch, a dielectric material and a ground plane fully covered with copper, as depicted in figure 3.24(a).

The resistive material has a relative permittivity of 3.4 and conductivity (σ) of 100 S m^{-1}, whereas the dielectric material has a relative permittivity of 3.5. The structure is ultrathin with the thickness on the order of $0.088\lambda_0$ and operates over 24.1–42.6 GHz. The absorber exhibits absorption above 90% over the 18.5 GHz bandwidth with peak absorption of 94% and 98% at 28.9 and 38 GHz, respectively, as shown in figure 3.24(b). It was observed by the researchers through electric field distribution that first resonance at 28.9 GHz occurs due to strong coupling between the FSS rings whereas the field is mostly concentrated at the patch for second resonance at 38 GHz. It was further investigated and realized that with the increase in conductivity of the resistive FSS the absorptivity gets reduced, whereas the dielectric loss has negligible contribution in the overall power consumption. The surface loss on the resistive FSS and distribution of power loss on the dielectric were observed as shown in figure 3.24(c), and it is evident that the surface loss is more significant.

Graphene-based frequency selective surfaces with cascaded metamaterial-inspired unit cells can be used as absorbers in the THz regime [58, 73]. The design methods are quite similar to that of the microwave absorbers with the difference that for THz absorbers the FSS layer can be cascaded with either a graphene monolayer or another graphene-based FSS. Mishra *et al* proposed such a THz absorber with the

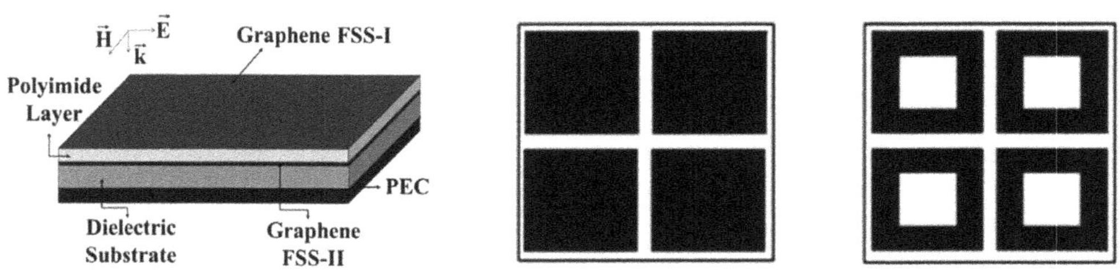

(a) (b) (c)

Figure 3.24. (a) MTM based Graphene FSS layered structure (b) square patch FSS (FSS-I) (c) square aperture FSS (FSS-II). Reproduced from [73]. CC BY 4.0.

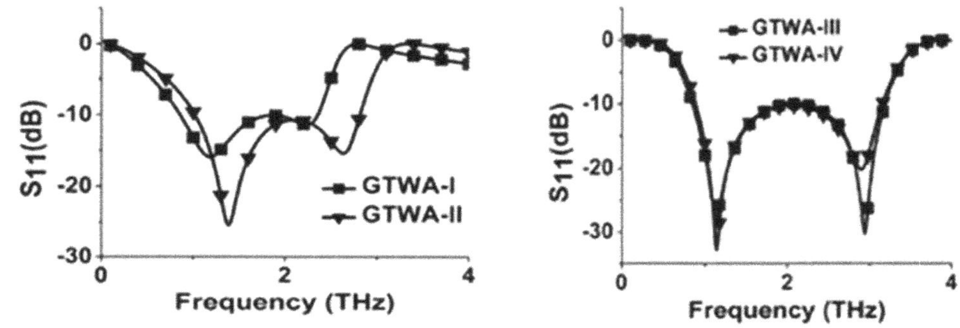

(a) (b)

Figure 3.25. Reflection coefficient (S11) for (a) GTWA-I and GTWA-II (b) GTWA-III and GTWA-IV. Reproduced from [73]. CC BY 4.0.

cascading of multiple layers, as shown in figure 3.24(a), where graphene-based FSS layers are separated by a thin polyimide layer that is further placed on a substrate ($\varepsilon_r = 5$) backed by a metallic ground plane (PEC) [73].

The researchers have used square patch-shaped FSS (SP-FSS) and square aperture-shaped FSS (SA-FSS) as shown in figures 3.24(b) and (c). They investigated four configurations with the combination of SP-FSS and graphene monolayer (GTWA-I), SA-FSS and graphene monolayer (GTWA-II), SP-FSS on the top and SA-FSS on the bottom (GTWA-III) and finally SA-FSS on the bottom and SP-FSS on the top (GTWA-IV). The reflection coefficient (S11) was analyzed for all the configurations as presented in figures 3.25(a) and (b) that reveal S11 below -10 dB over a wide bandwidth leading to wide band absorption. This is due to the fact that the absorption coefficient depends on both S11 and S21 (equation (3.12)), thus the presence of PEC ensures no transmission or S21 close to zero.

It was observed that with respect to -10 dB reference of S11, GTWA-I and GTWA-II exhibit absorption bandwidths of 1.50 and 1.79 THz, respectively. This reveals the advantage of using loop-shaped FSS over solid patch for absorber design. However with the hybrid configuration in GTWA-III and GTWA-IV the absorption bandwidth was further improved to 2.34 and 2.26 THz, respectively. Maximum absorption was achieved with the GTWA-IV showing peak S11 of −32.8 dB. Further, the response of the graphene-based absorber was studied against variation of chemical potential (μ_c) and it shows peak absorption at 2.34 THz for GTWA-III at μ_c of 0.38 and 0.25 eV. The absorption mechanism was investigated and it was found that impedance matching between input impedance of the structure and free space impedance is responsible for the absorption that is achieved by optimizing the unit cell dimensions through equivalent circuit modeling.

3.3.3 MTM-inspired polarization converter

Polarization conversion refers to the alteration of orientation of the linearly polarized signals from horizontal to vertical polarization or vice-versa, as well as switching the linear polarization to circular polarization or the reverse one. Polarization converters are used in radar cross-section reduction, satellite communication, etc. Traditional polarization converters for electromagnetic waves are designed using ferrite materials following the Faraday rotation leading to bulky structures. Metamaterial-inspired FSSs or metasurface-based polarization converters exhibit various kinds of polarization conversion including LP–LP, LP–CP, CP–LP as well as CP–CP using sub-wavelength and ultrathin structures [59, 60, 74, 75]. They can work either in transmission mode or reflection mode. In transmission mode the surface should exhibit minimal loss, whereas under reflection mode metal-backed dielectric structures are below the surface.

Dutta *et al* proposed a metasurface-based polarization converter that is capable of LP–LP conversion as well as LP–CP conversion over multiple frequency bands [74]. The unit cell based on metal-backed 3.2 mm thick FR4 substrate ($\varepsilon_r = 4.4$) consists of a diagonal strip and a meandered square ring. The square ring is loaded with a slit at both the corners parallel to the diagonal strip. The proposed MTM-based FSS-type absorber is reflective in nature, allowing negligible transmission of the microwaves.

A three-dimensional view of the array of a metasurface-based polarization converter exhibiting polarization conversion is shown in figure 3.26(a) along with the unit cell in figure 3.26(b). It was observed that any incident LP wave on the structure gets reflected with both co- and cross-polarized components because of the anisotropic nature of the unit cell. The reflected signal consists of the co-polarized component (R_{xx}) below −10 dB and the cross-polarized component (R_{yx}) close to −1 dB at 4.3, 7.2, 12.3 and 15.15 GHz. Moreover, the plot in figure 3.27 shows a phase difference of 0° or ±180° between R_{xx} and R_{yx} in the four bands that ensure linearly polarized reflected signals orthogonal to the incident signal. The polarization conversion ratio (PCR) was estimated using (3.13) and was found [74] to be in between 98% and 99% in all the four operating bands.

$$\mathrm{PCR} = \frac{R_{yx}^2}{R_{yx}^2 + R_{xx}^2} \tag{3.14}$$

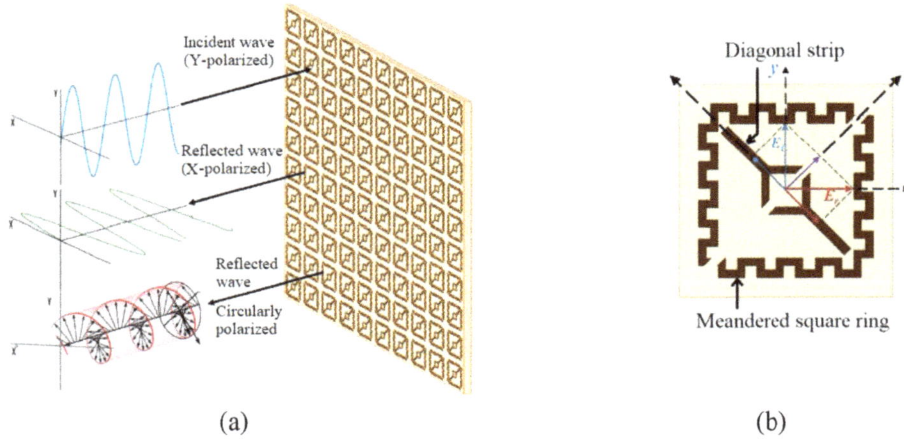

Figure 3.26. (a) Metasurface-based polarization converter for LP–LP as well as LP–CP conversion and (b) the unit cell of the MTM-inspired polarization converter. Reproduced from [74]. CC BY 4.0.

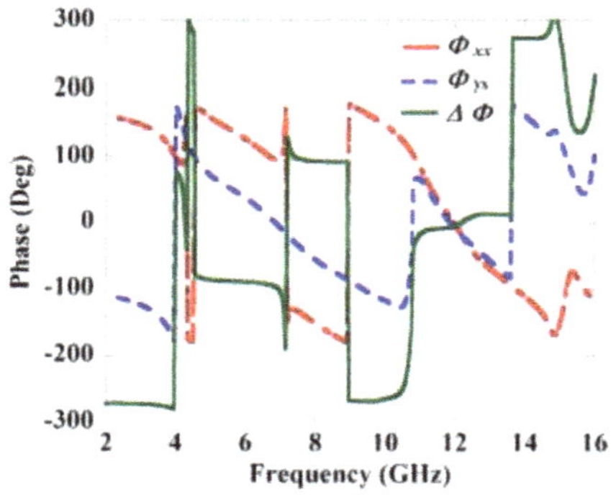

Figure 3.27. Phase response of the reflected signals. Reproduced from [74]. CC BY 4.0.

It was observed from the analysis of the co- and cross-polarized reflected waves in the operating bands of 4, 5.3, 8.5, and 14.5 GHz that the two orthogonally reflected waves have nearly equal amplitudes at the four frequency bands with a phase difference of ±180°. This ensures right-hand or left-hand circular polarization of the reflected signal at different frequencies. Figure 3.28 shows the surface current distribution on the unit cell that better justifies the polarization conversion.

It can be observed from figure 3.27(a) that the total current on the top layer or bottom layer of the MTM-inspired metasurface resulting from the addition of the orthogonal currents are parallel to each other at 4.3 GHz ensuring electric response of the structure. However, the total current on both the layers at 7.2 GHz, on the other hand, are opposite to each other, causing a circulating current that in turn generates a magnetic field leading to magnetic response of the structure. In both the cases the reflected signal is linearly polarized with the orthogonal counterpart. However, in the other four bands (LP–CP), for example at 5.3 GHz, the surface

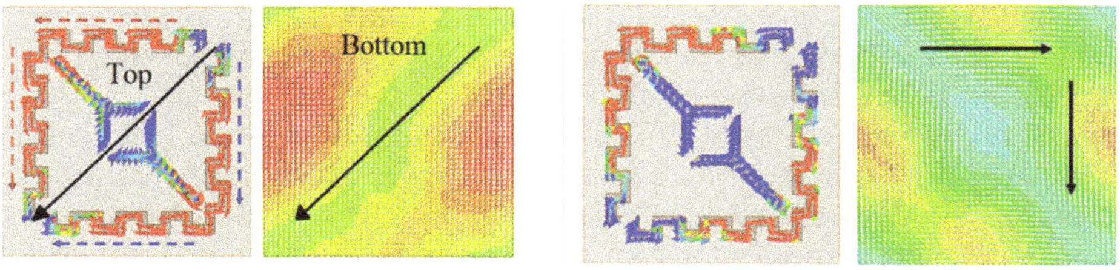

(a) 4.3 GHz (b) 5.3 GHz

Figure 3.28. Surface current distribution on the metasurface-based polarization converter. Reproduced from [74]. CC BY 4.0.

currents in the ground plane orient themselves in an orthogonal mode (figure 3.27 (b)) that indicates the circularly polarized reflected waves from the surface.

Polarization converters can also be realized in the THz regime using MTM-inspired periodic structures [60, 75]. Polarization converters are generally constructed by using metallic structures at microwave and millimeter-wave frequencies. However, fabrication challenges and large amounts of losses make these structures unsuitable at THz frequencies [75]. MTM-based polarimetric devices are chosen due to unavailability of suitable materials for terahertz device applications. Grady *et al* proposed polarization converters of reflection and transmission types using periodic structures based on MTMs [60]. The reflection type MTM consists of 200 nm thick gold cut-wire based on 33 μm thick polyimide dielectric backed by gold ground plane of 200 nm thickness, as shown in figure 3.28(a). The gold wire strips are placed at an angle of 45° with the x-axis. When the plane wave falls on the surface with the electric field oriented along the x-axis, it excites a dipolar oscillation $\boldsymbol{p}$ along the cut wires. The orizontal component of the oscillation p_x is responsible for co-polarized scattered fields, whereas the vertical component p_y results in cross-polarized scattering.

It was observed through the reflection coefficient of the polarizer that between 0.73 and 1.8 THz, the cross-polarized reflection is higher than 80% and the co-polarized reflection is mostly below 14%, leading to a reflected electric field along y-axis. A transmission type MTM-based polarizer is realized using orthogonally placed gold cut-wires on the two sides. In order to ensure maximum transmission the researchers replaced the metal backing (gold) with an orthogonal metal grating that also behaves like an isolator for the signals travelling towards the source. It was observed that the transmission type polarizer rotates the linear polarization by 90°, with a conversion efficiency of more than 50% from 0.52 to 1.82 THz.

3.3.4 MTM-based FSS for biomedical application

Medical telemetry is an integral part of biomedical engineering where patient's health parameters such as heart rate, blood pressure, blood glucose level, pulse rate, etc., are monitored remotely through sensors and antennas mounted on the body or clothes [76]. The system as a whole is known as a wireless body area network (WBAN) and it mostly operates in the lower end of the microwave spectrum, including ISM band frequencies [77]. The continuous microwave radiation from the wearable antenna may have detrimental effects on the human body, which is characterized by specific absorption rate (SAR) of the antenna that indicates the amount of radiation absorbed by the human body measured over 1 g of tissue [78]. MTM-based FSS structures with a reflective response can be integrated with these antennas to reduce their SAR level towards the body. On the other hand, wearable antennas generally possess poor gain that can even be negative at certain times (especially for implantable antennas where the antenna is mounted inside the body). In such cases MTM-based FSS structures with bandpass response can be integrated with such antennas to enhance the gain, but only in the broadside direction, leading to successful transmission and reception of the signals in the WBAN [79, 80].

Yalduz *et al* proposed a low profile wearable patch antenna with a metamaterial-based periodic structure as a reflector for low SAR, intended for medical telemetry applications over the ultrawide band of 4.5–13 GHz [79]. The patch and ground plane are shown in figure 3.29(a).

The metamaterial-inspired periodic structure or metasurface as shown in figure 3.29(b) consists of square patches with sub-wavelength periodicity on the order of $0.15\lambda_0$ that enables accommodation of a large number of unit cells beneath the antenna. The antenna, as well as the MTM, are based on a flexible felt substrate with the thickness of 1 mm for the antenna and 2 mm for the metasurface. This makes the overall wearable antenna dimension ultrathin with the thickness of 4.68 mm on the order of $\lambda_0/14$, suitable for mounting on wearable clothes or directly on the skin. During the analysis it was observed that the real part of the permittivity of the MTM-based FSS becomes negative below 6.8 GHz and above 11.5 GHz, whereas the permeability becomes negative between 6.8 and 11.5 GHz. This ensures a negative refractive index over the band.

The use of the metasurface below the antenna patch leads to a peak gain of 6 dBi and also reduces the front-to-back ratio (FBR) by 10–15 dB over the entire band, leading to significant reduction in back radiation towards the human body. The antenna-MTM performance was investigated by placing it above a human body

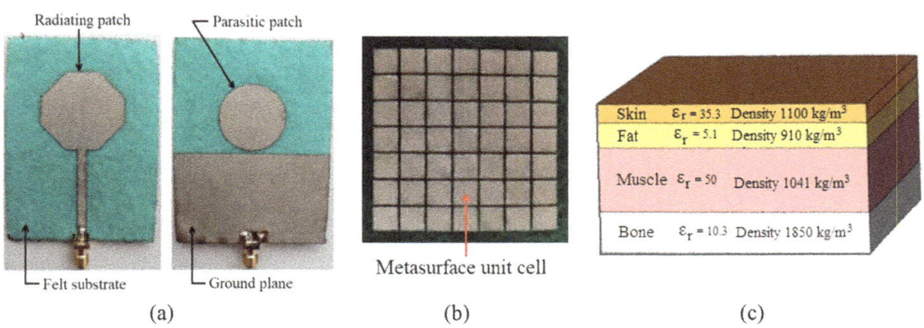

Figure 3.29. (a) Wearable antenna patch and ground plane on felt substrate, (b) MTM-based periodic structure as reflector, and (c) cubic human body tissue model with the properties at 4 GHz. Reprinted from [79], Copyright (2020), with permission from Elsevier.

tissue model as shown in figure 3.29(c) and the SAR was reduced from 6.27 to 0.067 W kg^{-1} over 10 g tissue at 4 GHz with a similar fall over the entire operating band (figure 3.30). This leads to a SAR reduction by 98%.

Metamaterial-inspired FSSs or metasurfaces can also be integrated with implantable antennas placed inside the human body beneath the skin [80, 81]. Implantable modules are applicable in prevention of strokes and heart attacks, monitoring blood glucose levels, functional electrical stimulators (FES), cochlear and retinal implants, etc. Metasurfaces can be used for SAR reduction of such antennas and broadside gain enhancement. Das *et al* proposed na MTM-based periodic structure with high absorptivity for the SAR reduction of a CPW-fed implantable antenna (figure 3.31(a)) at 2.4 GHz [80]. The design is composed of four Ω-shaped resonators placed diagonally and connected via X-shaped metallic strips, as shown in figure 3.31(b).

The Ω-shaped structure is used to realize bianisotropic properties. A flexible 0.5 mm thick PLA substrate with $\varepsilon_r = 2.72$ is used as substrate material for the MTM-based structure. The metasurface exhibits complex values of permittivity and

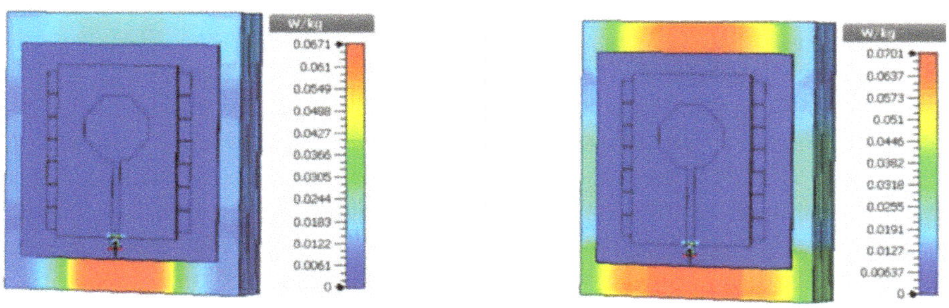

(a) 4 GHz (b) 10.5 GHz

Figure 3.30. Specific absorption rate (SAR) value of the antenna with the MTM. Reprinted from [79], Copyright (2020), with permission from Elsevier.

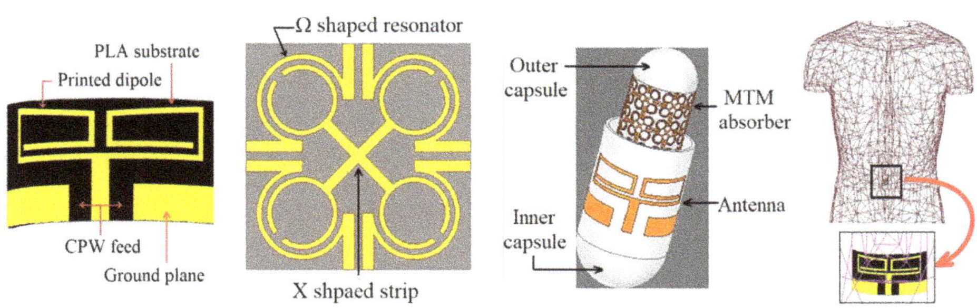

(a) (b) (c) (d)

Figure 3.31. (a) CPW-fed printed dipole antenna, (b) unit cell of the MTM-based FSS absorber, (c) capsule module integrated with antenna and MTM absorber, and (d) human torso model with the antenna. Reproduced from [80]. CC BY 4.0.

permeability that lead to attenuation of the waves radiated by antenna towards the MTM. Moreover the metasurface acts as a microwave absorber with high absorptivity even under bending conditions with radius of curvature between 10 and 100 mm that makes it suitable for reduction in antenna back radiation.

A capsule module of 9–10 mm diameter is used to accommodate the antenna and metasurface as shown with the simulation model in figure 3.31(c) and with the fabricated model in figure 3.32(b). The diameter of the capsule is suitable for passing through the gastrointestinal tract. The capsule is placed inside a human torso model, as shown in figure 3.31(d), for simulation and the fabricated prototype is placed in a human torso model (figure 3.32(a)) where the SAR value was investigated. With the metasurface the antenna SAR was reduced by 24% from 453 to 339 W kg^{-1}. Measurement of the antenna-MTM characteristics is done on a porcine slab as shown in figure 3.32(c) showing no significant change in antenna gain and SAR value.

Medical imaging plays an important role in biomedical engineering due to its use in early detection of fatal decease such as cancer [82]. Terahertz (THz) frequencies are very useful in medical imaging technology because of its low photon energy, causing non-ionizing characteristics as well as an ultra wideband spectrum comparable with molecular vibration resonances [82–84]. Due to these, THz signals are able to detect biological samples in label-free configurations. However, a significant amount of signals remain undetected due to water absorption, making it difficult to discriminate them at the THz regime. This problem is resolved by implementing THz imaging with enhanced color contrast by the use of localized and enhanced THz field with the incorporation of MTM-inspired nanometer-scale array structure [85]. Metamaterial-based periodic structures such as FSSs can produce a larger image contrast and enhanced sensitivity. THz MTM-based unit cells with sub-wavelength periodicity are much smaller compared to biological samples. Thus, they can be accommodated in a large number in the array structure, leading to better imaging of the biological samples in the THz regime.

Lee *et al* proposed the use of an MTM-inspired nano-slot array uniformly fabricated on a centimeter-scale for the imaging of bio-samples over a large-area [85]. The proposed nano-slot array, as shown in the figure 3.33(a), has horizontal

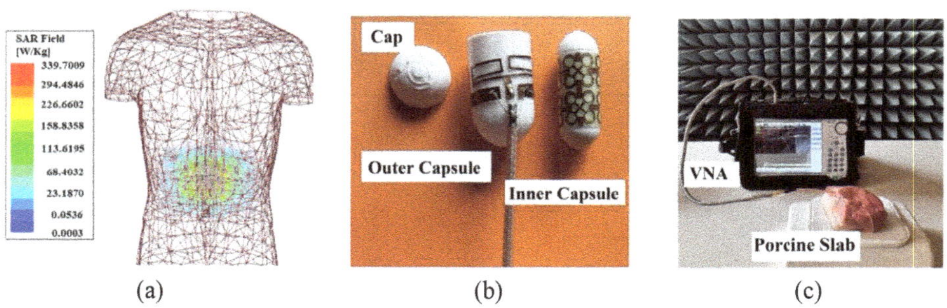

Figure 3.32. (a) SAR distribution in human torso with antenna, (b) prototype of the capsule with antenna and metasurface, and (c) antenna-MTM implanted in porcine slab for measurement. Reproduced from [80]. CC BY 4.0.

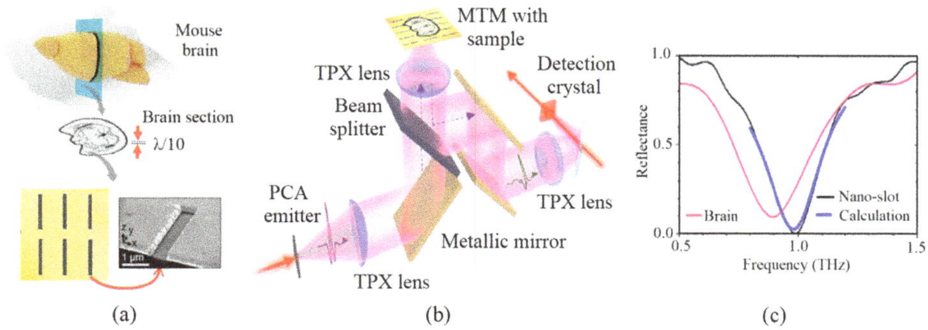

Figure 3.33. (a) Sectioned mouse brain tissue and nano-slot array with enlarged unit cell, (b) THz time-domain spectroscopy system, and (c) THz image of sagittal cross-section of mouse brain tissue. Reprinted from [85], Copyright (2020), with permission from Elsevier.

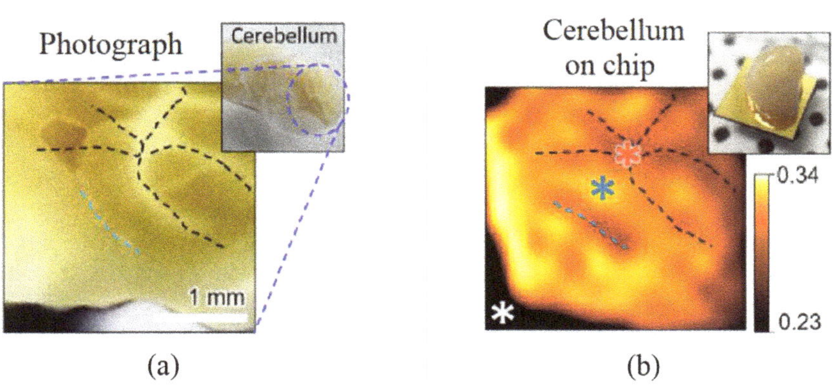

Figure 3.34. (a) Photographic image and (b) MTM-based nano-slot array driven THz image of sagittal cross-section of mouse brain tissue. Reprinted from [85], Copyright (2020), with permission from Elsevier.

and vertical periodicities of 40 and 10 μm, respectively, for each unit cell. The MTM-based array is fabricated on a 675 μm thick Si wafer with high resistivity ($\rho > 10\,000$ ohm-cm) to minimize the absorption loss. The nano-slots are embedded in a gold based metallic screen of 150 nm thickness chosen greater than the skin depth at 1 THz. A section of mouse brain tissue with dimensions of 2 cm × 2 cm and thickness of 30 μm ($\lambda/10$ at 1 THz) is placed above the MTM nano-slot loaded wafer of the same size for the imaging. The brain tissue loaded nano-slot array is investigated using a THz time-domain spectroscopy system as shown in figure 3.33(b). The reflectance of the MTM array was studied and shows a resonance close to 1 THz for a bare nano-slot array, model-suggested spectrum. The resonance was observed to be shifted in the presence of the brain tissue sample.

A wet sagittal section from a mouse cerebellum was tested using the THz spectroscopy and the THz image was compared with the photographic image. As can be observed from the photographic image in figure 3.34(a), the cerebellar cortex exhibits a darker color compared to the inner region of the cerebellum, which in turn symbolizes gray matter including cell bodies and capillaries. The arborvitae, on the other hand, appears white due to the presence of the myelin sheath encircled by nerve fibers. However, it is very difficult to distinguish between the white and gray matter if the biological sample is wet, due to water absorption and resulting interference. On the contrary, the THz image achieved at 0.95 THz clearly differentiates between the two areas, as shown in figure 3.34(b), using high contrast in the image. A difference of 4% was observed in the reflectance response between the white matter and gray matter.

References

[1] Gao B, Yuen M M and Ye T T 2017 Flexible frequency selective metamaterials for microwave applications *Sci. Rep.* **7** 45108

[2] Whitesides G M 2002 Organic materials science *MRS Bull.* **27** 56–65

[3] Marqués R, Martin F and Sorolla M 2011 *Metamaterials with Negative Parameters: Theory, Design, and Microwave Applications* (New York: Wiley)

[4] James J R and Hall P S 1989 *Handbook of Microstrip Antennas* (Stevenage: IET)

[5] Khan M U, Sharawi M S and Mittra R 2015 Microstrip patch antenna miniaturisation techniques: a review *IET Microw. Antennas Propag.* **9** 913–22

[6] Jamshed M A, Nauman A, Abbasi M A and Kim S W 2020 Antenna selection and designing for THz applications: suitability and performance evaluation: a survey *IEEE Access* **8** 113246–61

[7] Varamini G, Keshtkar A and Naser-Moghadasi M 2018 Compact and miniaturized microstrip antenna based on fractal and metamaterial loads with reconfigurable qualification *AEU-Int. J. Electron.* C **83** 213–21

[8] Yuan B, Zheng Y H, Zhang X H, You B and Luo G Q 2017 A bandwidth and gain enhancement for microstrip antenna based on metamaterial *Microw. Opt. Technol. Lett.* **59** 3088–93

[9] Wu W, Yuan B, Guan B and Xiang T 2017 A bandwidth enhancement for metamaterial microstrip antenna *Microw. Opt. Technol. Lett.* **59** 3076–82

[10] Lv Z, Li Z, Han Y, Cao Y and Yang L 2022 A wideband and ultra-thin metamaterial absorber based on resistive FSS *Symmetry* **14** 1148

[11] Li W, Lan Y, Wang H and Xu Y 2021 Microwave polarizer based on complementary split ring resonators frequency-selective surface for conformal application *IEEE Access* **9** 111383–9

[12] Jyothi M H, Bipin D V, Choudhury B and Nair R U 2016 Design of conformal metamaterial unit cells for invisibility cloaking applications *2016 IEEE Annual India Conf. (INDICON)* (Piscataway, NJ: IEEE) 1–5

[13] Smith D R and Pendry J B 2006 Homogenization of metamaterials by field averaging *JOSA* B **23** 391–403

[14] Smith D R, Padilla W J, Vier D C, Nemat-Nasser S C and Schultz S 2000 Composite medium with simultaneously negative permeability and permittivity *Phys. Rev. Lett.* **84** 4184

[15] Veselago V, Braginsky L, Shklover V and Hafner C 2006 Negative refractive index materials *J. Comput. Theor. Nanosci.* **3** 189–218

[16] Pendry J B, Holden A J, Robbins D J and Stewart W J 1998 Low frequency plasmons in thin-wire structures *J. Phys. Condens. Matter* **10** 4785

[17] Zhou J, Koschny T, Kafesaki M, Economou E N, Pendry J B and Soukoulis C M 2005 Saturation of the magnetic response of split-ring resonators at optical frequencies *Phys. Rev. Lett.* **95** 223902
[18] Bayatpur F 2009 Metamaterial-inspired frequency-selective surfaces Doctoral Dissertation, the University of Michigan
[19] Jang Y and Ahn D 2020 Wideband 90 phase shifter using modified composite resonant circuits for phase slope alignment *Microw. Opt. Technol. Lett.* **62** 1498–502
[20] Kiem N K, Phuong H N B, Hieu Q N and Chien D N 2015 A novel metamaterial MIMO antenna with high isolation for WLAN applications *Int. J. Antennas Propag.* **2015** 851904
[21] Mehdipour A, Denidni T A and Sebak A R 2013 Multi-band miniaturized antenna loaded by ZOR and CSRR metamaterial structures with monopolar radiation pattern *IEEE Trans. Antennas Propag.* **62** 555–62
[22] Tadesse A D, Acharya O P and Sahu S 2020 Application of metamaterials for performance enhancement of planar antennas: a review *Int. J. RF Microw. C.* E **30** e22154
[23] Hafez H A, Chai X, Ibrahim A, Mondal S, Férachou D, Ropagnol X and Ozaki T 2016 Intense terahertz radiation and their applications *J. Opt.* **18** 093004
[24] Hillger P, Grzyb J, Jain R and Pfeiffer U R 2018 Terahertz imaging and sensing applications with silicon-based technologies *IEEE Trans. Terahertz Sci. Technol.* **9** 1–9
[25] Milias C, Andersen R B, Lazaridis P I, Zaharis Z D, Muhammad B, Kristensen J T, Mihovska A and Hermansen D D 2021 Metamaterial-inspired antennas: a review of the state of the art and future design challenges *IEEE Access* **9** 89846–65
[26] Ameen M and Chaudhary R K 2020 Electrically small circularly polarized antenna using vialess CRLH-TL and fractals for L-band mobile satellite applications *Microw. Opt. Technol. Lett.* **62** 1686–96
[27] Raja K B and Pandian S C 2022 Low-profile metamaterial-based T-shaped engraved electrically small antenna design with wideband operating capability for WLAN/5G applications *Physica* B **646** 414359
[28] Huang C, Jiao Y C, Weng Z B and Li X 2018 A planar multiband antenna based on CRLH-TL ZOR for 4G compact mobile terminal applications *2018 Int. Workshop on Antenna Technology (iWAT)* (Piscataway, NJ: IEEE) 1–3
[29] Baek J, Lee Y and Choi J 2013 A wideband zeroth-order resonance antenna for wireless body area network applications *IEICE Trans. Commun.* **96** 2348–54
[30] Geetharamani G and Aathmanesan T 2020 Design of metamaterial antenna for 2.4 GHz WiFi applications *Wirel. Pers. Commun.* **113** 2289–300
[31] Rao M V, Madhav B T, Anilkumar T and Nadh B P 2018 Metamaterial inspired quad band circularly polarized antenna for WLAN/ISM/Bluetooth/WiMAX and satellite communication applications *AEU-Int. J. Electron.* C **97** 229–41
[32] Bickford J A, Duwel A E, Weinberg M S, McNabb R S, Freeman D K and Ward P A 2019 Performance of electrically small conventional and mechanical antennas *IEEE Trans. Antennas Propag.* **67** 2209–23
[33] Wu Z, Tang M C, Li M and Ziolkowski R W 2019 Ultralow-profile, electrically small, pattern-reconfigurable metamaterial-inspired Huygens dipole antenna *IEEE Trans. Antennas Propag.* **68** 1238–48
[34] Yang X, Xiao J, Wang J and Sheng L 2019 Compact frequency reconfigurable antennas based on composite right/left-handed transmission line *IEEE Access* **7** 131663–71

[35] Xie Z, Li Y, Sun L, Wu W, Cao R and Tao X 2020 A simple high-resolution near-field probe for microwave non-destructive test and imaging *Sensors* **20** 2670
[36] Catapano I, Gennarelli G, Ludeno G, Noviello C, Esposito G and Soldovieri F 2021 Contactless ground penetrating radar imaging: state of the art, challenges, and microwave tomography-based data processing *IEEE Geosci. Remote Sens. Mag.* **10** 251–73
[37] Lin X, Chen Y, Gong Z, Seet B C, Huang L and Lu Y 2020 Ultrawideband textile antenna for wearable microwave medical imaging applications *IEEE Trans. Antennas Propag.* **68** 4238–49
[38] Fiser O, Hruby V, Vrba J, Drizdal T, Tesarik J, Vrba Jr J and Vrba D 2022 UWB bowtie antenna for medical microwave imaging applications *IEEE Trans. Antennas Propag.* **70** 5357–72
[39] Ali W A, Mohamed H A, Ibrahim A A and Hamdalla M Z 2019 Gain improvement of tunable band-notched UWB antenna using metamaterial lens for high speed wireless communications *Microsyst. Technol.* **25** 4111–7
[40] Kundu S and Chatterjee A 2021 Sharp triple-notched ultra wideband Antenna with gain augmentation using FSS for ground penetrating radar *Wirel. Pers. Commun.* **117** 1399–418
[41] Gopikrishnan G, Akhter Z, Bhadouria V S and Akhtar M J 2020 Microwave imaging of hidden crevasse in glacier terrain using metamaterial loaded eight-slot Vivaldi antenna *J. Electromagn. Waves Appl.* **34** 259–74
[42] Pozar David M 2009 *Microwave Engineering* (New York: Wiley)
[43] Borja B, Tirado-Méndez J A and Jardon-Aguilar H 2018 An overview of UWB antennas for microwave imaging systems for cancer detection purposes *Prog. Electromagn. Res.* B **80** 173–98
[44] Scapaticci R, Di Donato L, Catapano I and Crocco L 2012 A feasibility study on microwave imaging for brain stroke monitoring *Prog. Electromagn. Res.* B **40** 305–24
[45] Alibakhshikenari M *et al* 2020 Metamaterial-inspired antenna array for application in microwave breast imaging systems for tumor detection *IEEE Access* **8** 174667–78
[46] Esfandiyari M, Lalbakhsh A, Jarchi S, Ghaffari-Miab M, Mahtaj H N and Simorangkir R B 2022 Tunable terahertz filter/antenna-sensor using graphene-based metamaterials *Mater. Des.* **220** 110855
[47] Amanatiadis S A, Karamanos T D and Kantartzis N V 2017 Radiation efficiency enhancement of graphene THz antennas utilizing metamaterial substrates *IEEE Antennas Wirel. Propag. Lett.* **16** 2054–7
[48] Luo Y, Zeng Q, Yan X, Jiang T, Yang R, Wang J, Wu Y, Lu Q and Zhang X 2019 A graphene-based tunable negative refractive index metamaterial and its application in dynamic beam-tilting terahertz antenna *Microw. Opt. Technol. Lett.* **61** 2766–72
[49] Geim A K and Novoselov K S 2007 The rise of graphene *Nat. Mater.* **6** 183–91
[50] Keshavarz A and Vafapour Z 2018 Water-based terahertz metamaterial for skin cancer detection application *IEEE Sens. J.* **19** 1519–24
[51] Sugapriya K and Omkumar S 2023 Textile UWB 5G antenna for human blood clot measurement *Intell. Autom. Soft Comput* **36** 803–18
[52] Geetharamani G and Aathmanesan T 2019 Metamaterial inspired THz antenna for breast cancer detection *SN Appl. Sci.* **1** 1–9
[53] Marqués R, Mesa F, Martel J and Medina F 2003 Comparative analysis of edge- and broadside-coupled split ring resonators for metamaterial design-theory and experiments *IEEE Trans. Antennas Propag.* **51** 2572–81

[54] Poorgholam-Khanjari S and Zarrabi F B 2021 Reconfigurable Vivaldi THz antenna based on graphene load as hyperbolic metamaterial for skin cancer spectroscopy *Opt. Commun.* **480** 126482
[55] Fang C Y, Gao J S and Liu H 2014 A novel metamaterial filter with stable passband performance based on frequency selective surface *AIP Adv.* **4** 077114
[56] Sorathiya V, Lavadiya S, Parmar B, Baxi S, Dhankot T, Faragallah O S, Eid M M and Rashed A N 2022 Tunable frequency selective surface using crossed shaped graphene metasurface geometry for far infrared frequency spectrum *Appl. Phys. B.* **128** 169
[57] Mishra N, Choudhary D K, Chowdhury R, Kumari K and Chaudhary R K 2017 An investigation on compact ultra-thin triple band polarization independent metamaterial absorber for microwave frequency applications *IEEE Access* **5** 4370–6
[58] Mishra R, Panwar R and Singh D 2018 Equivalent circuit model for the design of frequency-selective, terahertz-band, graphene-based metamaterial absorbers *IEEE Magn. Lett.* **9** 1–5
[59] Yan M, Wang J, Pang Y, Xu C, Chen H, Zheng L, Zhang J and Qu S 2019 An FSS-backed dual-band reflective polarization conversion metasurface *IEEE Access* **7** 104435–42
[60] Grady N K, Heyes J E, Chowdhury D R, Zeng Y, Reiten M T, Azad A K, Taylor A J, Dalvit D A and Chen H T 2013 Terahertz metamaterials for linear polarization conversion and anomalous refraction *Science* **340** 1304–7
[61] Narayan S, Gulati G, Sangeetha B and Nair R U 2018 Novel metamaterial-element-based FSS for airborne radome applications *IEEE Trans. Antennas Propag.* **66** 4695–707
[62] Simruni M and Jam S 2019 Design of high gain, wideband microstrip resonant cavity antenna using FSS superstrate with equivalent circuit model *AEU-Int. J. Electron. Commun.* **112** 152935
[63] Chatterjee A and Parui S K 2015 Gain enhancement of a wide slot antenna using a second-order bandpass frequency selective surface *Radioengineering* **24** 455–61
[64] Al-Joumayly M and Behdad N 2009 A new technique for design of low-profile, second-order, bandpass frequency selective surfaces *IEEE Trans. Antennas Propag.* **57** 452–9
[65] Kou N, Liu H and Li L 2017 A transplantable frequency selective metasurface for high-order harmonic suppression *Appl. Sci.* **7** 1240
[66] Ramli A, Ismail A, Abdullah R S, Mahdi M A and Al-Hawari A R 2018 Miniaturize negative index metamaterial structure loaded filtenna *Prog. Electrom. Res.* M **72** 97–104
[67] Haghzadeh M and Akyurtlu A 2016 All-printed, flexible, reconfigurable frequency selective surfaces *J. Appl. Phys.* **120** 184901
[68] Ebrahimi A, Nirantar S, Withayachumnankul W, Bhaskaran M, Sriram S, Al-Sarawi S F and Abbott D 2015 Second-order terahertz bandpass frequency selective surface with miniaturized elements *IEEE Trans. Terahertz Sci. Technol.* **5** 761–9
[69] Xu W, Xie L and Ying Y 2017 Mechanisms and applications of terahertz metamaterial sensing: a review *Nanoscale* **9** 13864–78
[70] Nornikman H, Soh P J, Azremi A A, Husna M R and Liam O S 2008 Parametric study of pyramidal microwave absorber design *Int. Symp. Antennas Propag. (ISAP 2008)* **27** 30
[71] Sood D and Tripathi C C 2015 A wideband wide-angle ultra-thin metamaterial microwave absorber *Prog. Electrom. Res.* M **44** 39–46
[72] Singh A K, Abegaonkar M P and Koul S K 2018 Dual-and triple-band polarization insensitive ultrathin conformal metamaterial absorbers with wide angular stability *IEEE Trans. Electromagn. Compat.* **61** 878–86

[73] Mishra R, Sahu A and Panwar R 2019 Cascaded graphene frequency selective surface integrated tunable broadband terahertz metamaterial absorber *IEEE Photonics J.* **11** 1-0

[74] Dutta R, Ghosh J, Yang Z and Zhang X 2021 Multi-band multi-functional metasurface-based reflective polarization converter for linear and circular polarizations *IEEE Access* **9** 152738–48

[75] Cheng Z and Cheng Y 2019 A multi-functional polarization convertor based on chiral metamaterial for terahertz waves *Opt. Commun.* **435** 178–82

[76] Nikita K S (ed) 2014 *Handbook of Biomedical Telemetry* (New York: Wiley)

[77] Sánchez-Montero R, Camacho-Gómez C, López-Espí P L and Salcedo-Sanz S 2018 Optimal design of a planar textile antenna for industrial scientific medical (ISM) 2.4 GHz wireless body area networks (WBAN) with the CRO-SL algorithm *Sensors* **18** 1982

[78] Tuovinen T, Berg M and Salonen E T 2014 Antenna close to tissue: avoiding radiation pattern minima with an anisotropic substrate *IEEE Antennas Wirel. Propag. Lett.* **13** 1680–3

[79] Yalduz H, Tabaru T E, Kilic V T and Turkmen M 2020 Design and analysis of low profile and low SAR full-textile UWB wearable antenna with metamaterial for WBAN applications *AEU-Int J. Electron.* C **126** 153465

[80] Mitra S, Chezhian D, Mandal A S and Augustine B R 2022 A novel SAR reduction technique for implantable antenna using conformal absorber metasurface *Front. Med. Technol.* **4** 924433

[81] Mitra S, Mandal D and Augustine B R 2020 Implantable antenna gain enhancement using liquid metal-based reflector *Appl. Phys.* A **126** 1–7

[82] AlSawaftah N, El-Abed S, Dhou S and Zakaria A 2022 Microwave imaging for early breast cancer detection: current state, challenges, and future directions *J. Imaging* **8** 123

[83] Sun Q, He Y, Liu K, Fan S, Parrott E P and Pickwell-MacPherson E 2017 Recent advances in terahertz technology for biomedical applications *Quant Imaging Med Surg* **7** 345

[84] Peng Y, Shi C, Wu X, Zhu Y and Zhuang S 2020 Terahertz imaging and spectroscopy in cancer diagnostics: a technical review *BME Front.* **2020** 2547609

[85] Lee S H, Shin S, Roh Y, Oh S J, Lee S H, Song H S, Ryu Y S, Kim Y K and Seo M 2020 Label-free brain tissue imaging using large-area terahertz metamaterials *Biosens. Bioelectron.* **170** 112663

[86] Rao N N 2006 Elements of Engineering Electromagnetics, 6th edn (London: Pearson Education)

Chapter 4

Metamaterial inspired graphene-based antenna and FSS design

4.1 Introduction

Graphene has been termed as a wonder material since its discovery. Scientists have shown its application for diverse fields. It finds application in fields like super conductivity, bio sensor, photonics, water filtration, electromagnetic, defense armor, etc. It can also be used as metamaterials for designing antennas, electromagnetic absorbers and frequency selective surfaces (FSSs). A nano-dipole antenna structure using graphene as the metamaterial layer has been presented [1, 2]. In this chapter, the resonant frequency of the antenna structures, which lie in the visible spectrum (400–1000 nm), is controlled by using graphene as a material that will be transferred on top of the plasmonic antennas instead of creating plasmons in graphene. In this chapter, the response of the nano-dipole antenna structure, after the application of single and multiple graphene layer coatings, is examined. Generally, to analyze nano-antennas working in the optical spectrum, a standard dipole antenna structure is used [1, 2]. Various antenna parameters can be augmented with the use of graphene as a metasurface above the antenna patch. Multi-layered graphene-based antennas exhibit better performance compared to single-layered structures. A single-layered graphene-sheet resonant plasmon antenna is presented in [3] where a THz continuous-wave (CW) photomixer is used to feed the antenna, as depicted in figure 4.1(a). The photomixer excites the patch resonance, which enables radiation, as can be observed from near field distribution on the patch in figure 4.1(b). However, it exhibits moderately less directivity towards the substrate direction.

The use of a graphene metasurface as a superstrate above a patch antenna for the enhancement of multiple antenna parameters was implemented by a group of

doi:10.1088/978-0-7503-5422-6ch4

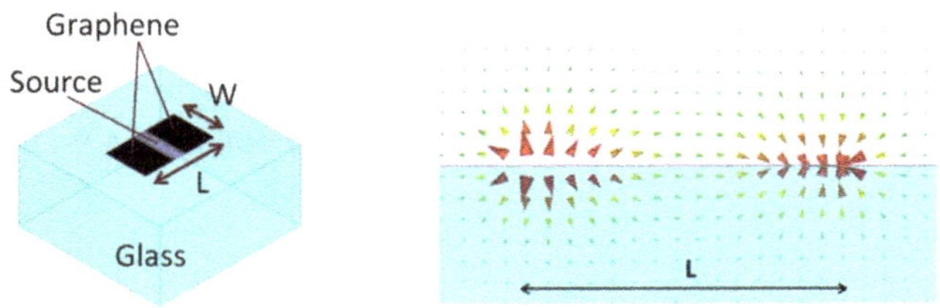

(a) (b)

Figure 4.1. (a) Single-layer graphene-based planar dipole antenna. (b) Electric near field distribution. Reprinted from [3], with the permission of AIP Publishing.

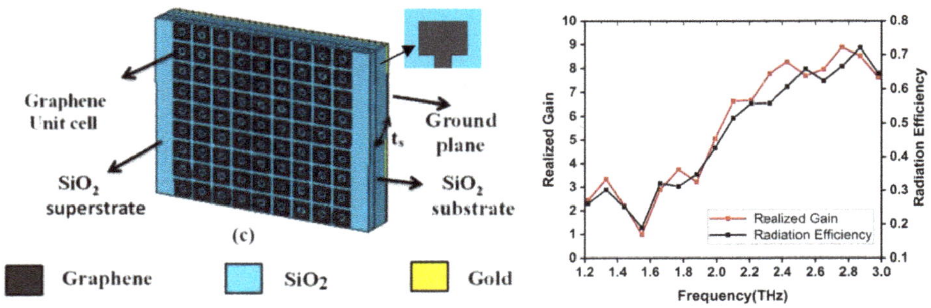

(a) (b)

Figure 4.2. (a) A patch antenna with the graphene-based metasurface as superstrate and (b) realized gain and efficiency of the antenna-metasurface. Reprinted from [4], Copyright (2022), with permission from Elsevier.

researchers [4]. As shown in figure 4.2(a), an array of graphene-based metasurfaces has been designed on a silicon dioxide substrate. The unit cell dimension is smaller compared to a quarter wavelength, maintaining the effective homogeneity limit. The antenna with the metasurface exhibits a fractional bandwidth of 69.15% and a return loss of 35.54 dB. Additionally, the graphene metasurface increases the directivity of the patch antenna. The realized gain of the antenna with the metasurface is enhanced to 8 dBi beyond 2.2 THz, as can be seen from figure 4.2(b).

4.2 Analysis of graphene-based metamaterials

In graphene, plasmonic oscillations are produced at mid- and far-infrared frequencies [1–5]. Hence, graphene is not plasmonic at optical frequencies and has a real part of the dielectric constant that is larger than zero at visible frequencies. Equation (4.1) provides the complex electrical permittivity for graphene [6, 7].

$$\varepsilon_{\text{eff}} = 1 - j\frac{\sigma_c}{\omega\varepsilon_0 d} = \left(1 + \frac{\sigma_i}{\omega\varepsilon_0 d}\right) - j\frac{\sigma_r}{\omega\varepsilon_0 d} \tag{4.1}$$

where σ_r and σ_i are computed using [2, 5–10], and σ_c is the optical conductivity. According to the ideas covered in [8], any material in its electrical equivalent circuit functions as a resistance if $\varepsilon_i \neq 0$ and as a capacitor if $\varepsilon_r > 0$, respectively, at a given frequency of incident radiation in the visible spectrum. Reflective spectroscopy has been used in experiments to verify graphene's refractive index in [9–11]. Thus, using the methods in [6, 9–11], it can be concluded that, at optical range, graphene functions as a capacitor and resistor combined, as shown in figure 4.3 [12].

The SEM images of figure 4.4 show that the size of the dipole nano-antennas has decreased [13]. Manufacturing the precise rectangle structure is difficult. The corners of the structures are curved. As a result, the predicted simulation response is blue shifted. The fabricated device's lower volume compared to the simulation's devices is the cause of the blue shift [14]. In order to reduce costs, a 5% tolerance was specified for the dipole nano-antenna's fabrication. The FDTD simulation of the scattering of the dipole nano-antenna fabrication with 5% tolerance was observed for verification.

As another example of graphene-based antennas, a tunable graphene-metal based antenna was studied [15]. This antenna is fed by a waveguide. The antenna is studied

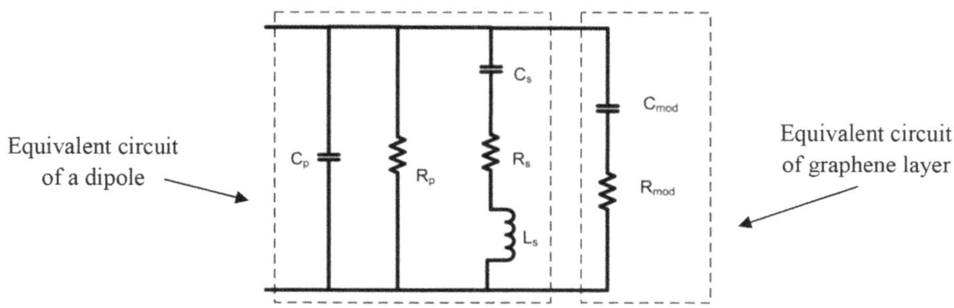

Figure 4.3. Equivalent circuit for the dipole nano-antenna covered with graphene. Graphene is represented as a combination of capacitor and resistor. Reproduced from [12]. CC BY 4.0.

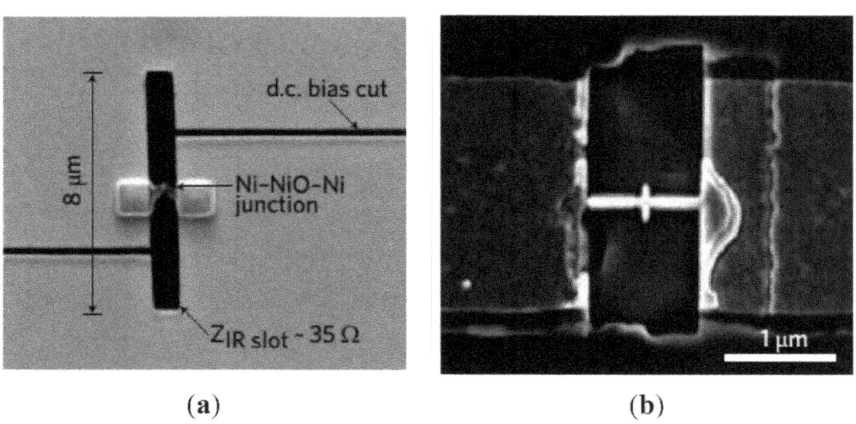

Figure 4.4. SEM images of the (a) slot antenna and (b) open-sleeve dipole antenna. Reproduced from [13]. CC BY 4.0.

using two different methods. First, by using numerical analysis; second, by transmission line model.

For the simulation, graphene is modeled using two methods. Firstly, graphene is regarded as an impedance sheet boundary condition, and secondly, graphene is represented by a thin bulk material layer. Even though the outcomes are almost identical, it is clear that two different issues are being handled. Overall, these findings support the effectiveness of both the analytical design approach and the recommended antenna. It is important to remember that the fabrication method affects graphene's ability to relax at the appropriate rate.

Lower values of τ typically have the effect of increasing losses, which lowers the antenna's efficiency. Moreover, it could be difficult to achieve high chemical potential values, especially for few-layer graphene. Generally, a smaller antenna frequency tuning corresponds to a decreased μc range. This chapter proposes a high radiation efficiency tunable waveguide-fed terahertz antenna. The propagation characteristics of the hybrid graphene-metal waveguide are investigated using the finite-element method as the numerical method and the EIM and transfer matrix theory as the analytical approaches. The analytical model is shown in figure 4.5.

An analytical formula for the antenna design is developed using the transmission line model, assuming quasi-TEM mode guidance in the proposed structure. The method that was developed is used to design a 2.7 THz waveguide-fed antenna. The final antenna structure is shown in figure 4.6. The suggested antenna has good impedance matching acceptable radiation efficiency and gain in the terahertz frequencies, according to the results of full-wave electromagnetic simulation. The antenna that is being presented can be used for a variety of radiated-wave applications, including sensing and wireless communication in electromagnetic nano networks.

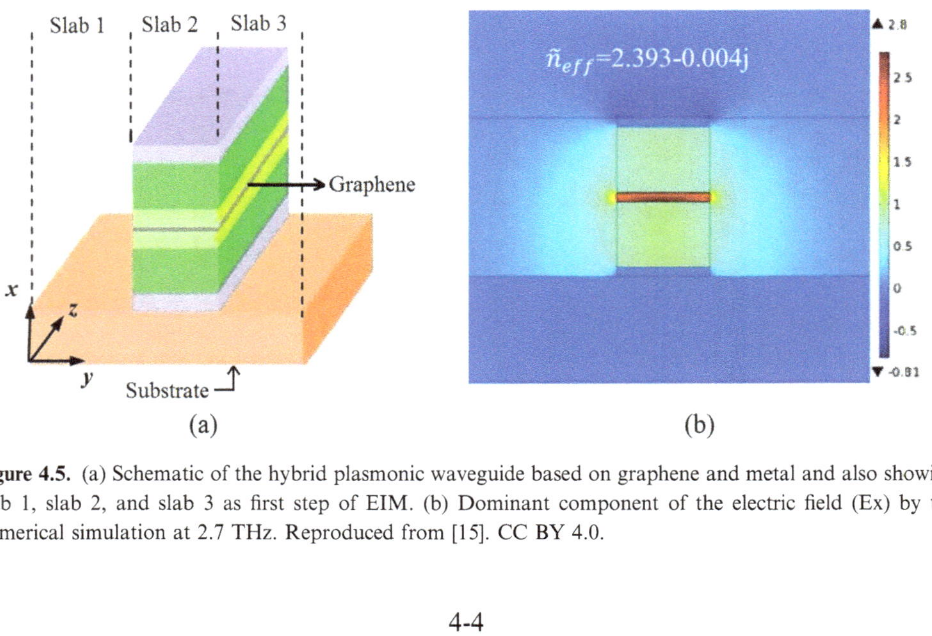

Figure 4.5. (a) Schematic of the hybrid plasmonic waveguide based on graphene and metal and also showing slab 1, slab 2, and slab 3 as first step of EIM. (b) Dominant component of the electric field (Ex) by the numerical simulation at 2.7 THz. Reproduced from [15]. CC BY 4.0.

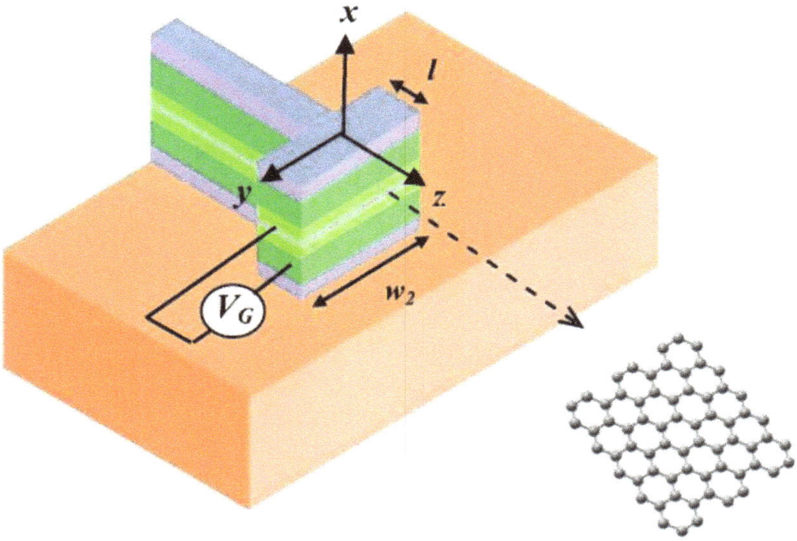

Figure 4.6. Schematic of the proposed terahertz antenna based on the hybrid graphene-metal structure. Reproduced from [15]. CC BY 4.0.

Antennas based on graphene are used in radar systems. In communication systems as well, antennas form essential components. By adjusting to the changing environmental conditions and offering more functionality for sending or receiving data, antennas' adjustability can improve the functionality of these systems. Antenna functionality can be altered by a variety of devices, including switches (PIN diodes, FETs), variable reactive loading, structural modifications, etc [16–18]. These techniques do have certain limitations, though, like a small dynamic range in variable reactive loading or a high tuning speed in PIN diodes. Re-configurability in terms of illumination direction, polarization, bandwidth and center frequency, and other aspects can be achieved by combining graphene with metamaterials and metasurfaces [19–22].

Graphene-based metamaterials are employed not only in antenna configuration but also in other fields, like the generation of orbital angular momentum waves, and development of linear to circular polarization converters [23, 24].

The ability to enable a broad range of re-configurability characteristics in graphene-based antennas [25, 26] is the biggest benefit of graphene metamaterials integration. In the GHz band, Huang *et al* developed a reconfigurable antenna using graphene [27], modeling graphene as a two-dimensional layer with chemical potential-dependent surface impedance (Rs). For $Rs = 2000\ \Omega$, this graphene array antenna exhibits an insertion loss of 9.55 dB at 7.89 GHz. Nevertheless, the reflection decreases to 1.1 dB when the graphene's surface impedance reaches $Rs = 100\ \Omega$. An array antenna with a right/left-handed feeding network was first introduced in [28]. The operating frequency of the antenna is reconfigurable due to the presence of a graphene layer beneath this feeding network. The operating frequency is controlled by variations in chemical potential between 0 and 0.2 eV.

Furthermore, a change in the antenna's gain was brought about by the variation in chemical potential. In [29], a circular horn antenna was used to feed graphene metamaterials for the design of an array antenna. The frequency range of 400–620 GHz yields negative values for er and lr for various chemical potential values. In this chapter, the highest gain of 22.6 dB was achieved at a chemical potential of 0.4 eV. Another instance of frequency modification with graphene metamaterials was demonstrated in [22], wherein the resonance frequency was adjusted by a graphene-based array of split-ring resonators placed beneath the main antenna.

Because of the growing applications related to IoT and non-geostationary satellite communications, the ability to reconfigure radiation patterns is becoming one of the most important characteristics of contemporary antennas [30–34]. The ability to increase channel capacity is greatly aided by this feature [35]. Esfandiari *et al* created a 2 × 2 MIMO antenna by rotating the antenna's pattern between 0° and 37° using an upper layer positioned with space above the antenna, as seen in figure 4.7(a) [36]. Figure 4.7(b). By creating a scenario as shown in figure 4.7(c), the channel's capacity for various transmitted powers was determined.

This accomplishment showed that aligning the pattern with the direction of the minimum loss increases the channel capacity by 47%. In [37], metal and graphene were used to produce a negative refractive index. This metamaterial structure altered the chemical potential of graphene, causing the designed antenna to rotate its pattern between 0 °and 27°. An effective method for building antennas with reversible patterns is this design approach.

As reported in [38, 39], the application of metallic metamaterials over graphene layers produced a 360° antenna pattern. By varying the chemical potential from 0 to 0.5 eV, the graphene/gold-based unit cells were able to achieve high transmission (ON state) and nearly total reflection (OFF state) at two frequency ranges independently, which caused the main antenna beam to steer. Six screen components made of graphene and gold metamaterials enclose the main antennas. Another illustration of how graphene metasurfaces can be used to modify the pattern of a leaky-wave antenna is shown in [40], where the periodic gaps created between the layers of the graphene metasurfaces are comparable to a row of capacitors facing the direction of illumination.

A minimum beam steering angle of 29° at 1 THz and a maximum angle of 75° at 3.41 THz have been attained with this technique. In addition to the methods for adjusting radiation patterns mentioned above, Lalbakhsh *et al* have created a novel class of all-metal metasurfaces with unmatched electromagnetic properties, a breakthrough in antenna engineering [41].

Figure 4.8 depicts the antenna's construction as well as the unit cell. Applying the period boundary conditions in the simulation is necessary when computing the scattering parameters of unit cells numerically using the configuration described in numerous computational techniques, such as finite-element, finite-difference time-domain, and finite-difference frequency-domain are used to numerically calculate the frequency response of electromagnetic components, such as antennas, sensors, filters, couplers, absorbers, and photonic crystals.

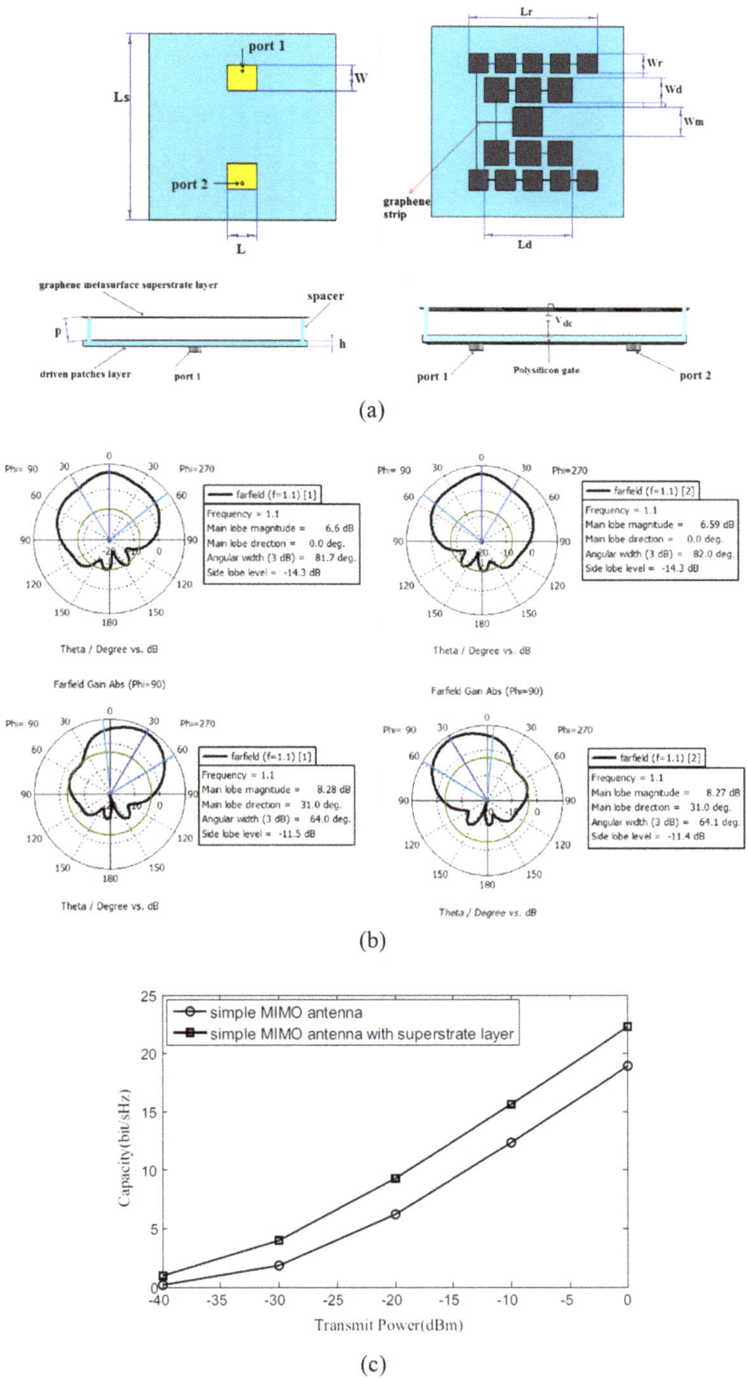

Figure 4.7. (a) The MIMO antenna's dimensions and geometry. (b) The antenna's radiation pattern with the superstrate layer in place and $\mu_c = 0$ eV for port 1, $\mu_c = 0$ eV for port 2, $\mu_c = 1$ eV for port 1, and $\mu_c = 1$ eV for port 2. (c) The designed scenario's channel capacity with an 80 dBm noise power. Reprinted by permission from Springer Nature Customer Service Centre GmbH: [Nature][Optical and Quantum Electronics] [36], Copyright (2019).

Figure 4.8. The suggested metasurface's structure and unit cell. Reprinted with permission from [42]. © The Optical Society.

References

[1] Fischer H and Martin O J F 2008 Engineering the optical response of plasmonic nano-antennas *Opt. Exp.* **16** 9144–54

[2] Ju L *et al* 2011 Graphene plasmonics for tunable terahertz metamaterials *Nat. Nanotechnol.* **6** 630–4

[3] Tamagnone M, Gomez-Diaz J S, Mosig J R and Perruisseau-Carrier J 2012 Analysis and design of terahertz antennas based on plasmonic resonant graphene sheets *J. Appl. Phys.* **112**

[4] Shubham A, Samantaray D, Ghosh S K, Dwivedi S and Bhattacharyya S 2022 Performance improvement of a graphene patch antenna using metasurface for THz applications *Optik* **264** 169412

[5] Yan H, Li X, Chandra B, Tulevski G, Wu Y, Freitag M, Zhu W, Avouris P and Xia F 2012 Tunable infrared plasmonic devices using graphene/insulator stacks *Nat. Nanotechnol.* **7** 330–4

[6] Simsek E 2013 A closed-form approximate expression for the optical conductivity of graphene *Opt. Lett.* **38** 1437–9

[7] Simsek E 2013 Improving tuning range and sensitivity of localized SPR sensors with graphene *IEEE Photon. Technol. Lett.* **25** 867–70

[8] Engheta N, Salandrino A and Alu A 2005 Circuit elements at optical frequencies: nanoinductors, nanocapacitors, and nanoresistors *Phys. Rev. Lett.* **95** 095504-1–094

[9] Ghamsari B G, Tosado J, Yamamoto M, Fuhrer M S and Anlage S M 2012 *Determination of the optical index for few-layer graphene by reflectivity spectroscopy* arXiv preprint arXiv:1210.0575.

[10] Weber J, Calado V and Van de Sanden M 2010 Optical constants of graphene measured by spectroscopic ellipsometry *Appl. Phys. Lett.* **97** 091904-1–3

[11] Bruna M and Borini S 2009 Optical constants of graphene layers in the visible range *Appl. Phys. Lett.* **94** 031901-1–033

[12] Mehta B, Benkstein K, Semancik S *et al* 2016 Gas sensing with bare and graphene-covered optical nano-antenna structures *Sci. Rep.* **6** 21287

[13] Kausar A S M Z, Reza A W, Latef T A, Ullah M H and Karim M E 2015 Optical nano antennas: state of the art, scope and challenges as a biosensor along with human exposure to nano-toxicology *Sensors* **15** 8787–831

[14] de Arquer F P G, Volski V, Verellen N, Vandenbosch G A E and Moshchalkov V V 2011 Engineering the input impedance of optical nano dipole antennas: materials, geometry and excitation effect *IEEE Trans. Antennas Propag.* **59** 3144–53 http://ieeexplore.ieee.org/lpdocs/epic03/wrapper.htm?arnumber=5948358
[15] Ullah Z, Witjaksono G, Nawi I, Tansu N, Irfan Khattak M and Junaid M 2020 A review on the development of tunable graphene nanoantennas for terahertz optoelectronic and plasmonic applications *Sensors* **20** 1401
[16] Lalbakhsh A, Simorangkir R B, Bayat-Makou N, Kishk A A and Esselle K P 2022 Advancements and artificial intelligence approaches in antennas for environmental sensing *Artificial Intelligence and Data Science in Environmental Sensing* (Berlin: Springer) 19–38
[17] Adibi S, Honarvar M A and Lalbakhsh A 2021 Gain enhancement of wideband circularly polarized UWB antenna using FSS *Radio Sci.* **56** e2020RS007098
[18] Das P, Mandal K and Lalbakhsh A 2022 Beam-steering of microstrip antenna using single-layer FSS based phase-shifting surface *Int. J. RF Microw. Comput. Aided Eng.* **32** e23033
[19] Jornet J M and Akyildiz I F 2013 Graphene-based plasmonic nano-antenna for terahertz band communication in nanonetworks *IEEE J. Sel. Areas Commun.* **31** 685–94
[20] Degl'Innocenti R *et al* 2014 Graphene-based optical modulator realized in metamaterial split-ring resonators operating in the THz frequency range *Terahertz, RF, Millimeter, and Submillimeter-Wave Technology and Applications VII* Vol. 8985 (Bellingham, WA: International Society for Optics and Photonics) p. 89851E
[21] Zhao X, Yuan C, Zhu L and Yao J 2016 Graphene-based tunable terahertz plasmoninduced transparency metamaterial *Nanoscale* **8** 15273–80
[22] Radwan A, D'Amico M, Din J, Gentili G G and Verri V 2016 Bandwidth and gain enhancement of a graphene-based metamaterial antenna for the THz band *ARPN J. Eng. Appl. Sci.* **11** 6349–54
[23] Cheng Y and Wang J 2021 Tunable terahertz circular polarization convertor based on graphene metamaterial *Diam. Relat. Mater.* **119** 108559
[24] Quader S, Zhang J, Akram M R and Zhu W 2020 Graphene-based high-efficiency broadband tunable linear-to-circular polarization converter for terahertz waves *IEEE J. Sel. Top. Quantum Electron.* **26** 1–8
[25] Mehta B and Zaghloul M E 2014 Tuning the scattering response of the optical nano antennas using graphene in *IEEE Photonics J.* **6** 1–8 Art no. 4800208
[26] Koppens F H, Chang D E and Garcia de Abajo F J 2011 Graphene plasmonics: a platform for strong light–matter interactions *Nano Lett.* **11** 3370–7
[27] Huang X *et al* 2016 Graphene metamaterials array based reconfigurable antenna *2016 Int. Symp. on Antennas and Propagation (ISAP)* (Piscataway, NJ: IEEE) 106–7
[28] Abdel Aziz A A, Ibrahim A A and Abdalla M A 2019 Tunable array antenna with CRLH feeding network based on graphene *IETE J. Res.* **68** 1713–21
[29] Zainud-Deen S H, Mabrouk A M and Malhat H 2018 Terahertz graphene based metamaterial transmitarray *Wirel. Pers. Commun.* **100** 1235–48
[30] Afzal M U, Esselle K P and Lalbakhsh A 2018 A methodology to design a low-profile composite-dielectric phase-correcting structure *IEEE Antennas Wirel. Propag. Lett.* **17** 1223–7
[31] Afzal M U, Esselle K P and Lalbakhsh A 2018 A metasurface to focus antenna beam at offset angle *2018 2nd URSI Atlantic Radio Science Meeting (ATRASC)* 28 May–1 June 2018 1–4

[32] Afzal M U, Matekovits L, Esselle K P and Lalbakhsh A 2020 Beam-scanning antenna based on near-electric field phase transformation and refraction of electromagnetic wave through dielectric structures *IEEE Access* **8** 199242–53

[33] Lalbakhsh A, Afzal M U, Esselle K P and Smith S L 2019 Wideband near-field correction of a fabry–perot resonator antenna *IEEE Trans. Antennas Propag.* **67** 1975–80

[34] Lalbakhsh A, Afzal M U, Esselle K P and Smith S L 2020 Low-cost nonuniform metallic lattice for rectifying aperture near-field of electromagnetic bandgap resonator antennas *IEEE Trans. Antennas Propag.* **68** 3328–35

[35] Jarchi S 2021 Radiation pattern direction control of THz antenna with applying planar graphene metasurface *Optik* **243** 167458

[36] Esfandiyari M, Jarchi S and Ghaffari-Miab M 2019 Channel capacity enhancement by adjustable graphene-based MIMO antenna in THz band *Opt. Quant. Electron.* **51** 1–11

[37] Luo Y *et al* 2019 A graphene-based tunable negative refractive index metamaterial and its application in dynamic beam-tilting terahertz antenna *Microwave Opt. Technol. Lett.* **61** 2766–72

[38] Yang Y-J, Wu B and Zhao Y-T 2021 Dual-band beam steering THz antenna using active frequency selective surface based on graphene *EPJ Appl. Metamater.* **8** 12

[39] Wu B, Hu Y, Zhao Y T, Lu W B and Zhang W 2018 Large angle beam steering THz antenna using active frequency selective surface based on hybrid graphenegold structure *Opt. Express* **26** 15353–61

[40] Gao M, Li K, Kong F, Zhuang H and Zhu G 2020 Graphene-based composite right/left-handed leaky-wave antenna at terahertz *Plasmonics* **15** 1199–204

[41] Lalbakhsh A, Afzal M U, Esselle K P and Smith S L 2022 All-metal wideband frequency-selective surface bandpass filter for TE and TM polarizations *IEEE Trans. Antennas Propag.* **70** 2790–800

[42] Wang Y, Wang Y, Li Q, Zhang Y, Yan S and Wang C 2021 Tunable graphene-based metasurface for an ultra-low sidelobe terahertz phased array antenna *Opt. Express* **29** 26865–75

Chapter 5

Design of metamaterial-inspired antennas and FSSs for conformal applications

This chapter focuses primarily on various aspects of antennas and frequency selective surfaces (FSSs) with metamaterial-inspired unit cells based on ultrathin substrates for conformal applications. The design methodologies of metamaterial-inspired antennas and FSS structures based on conformal substrates are also highlighted from an engineering point of view in this chapter, followed by discussion on various challenges in the design processes. A review of the state-of-the-art metamaterial-inspired antennas and FSS structures, especially for ultrathin non-planar designs, is included. An update on the latest technologies developed for such structures is provided following the introduction. The innovative technology that has been adopted to miniaturize the dimensions of the metamaterial-inspired FSSs in order to fit for curved structures is introduced accordingly. Based on the technology of cascading multiple layers of periodic structures, compact FSS designs in terms of wavelength corresponding to the operating frequency are developed. Finally, incorporation of metamaterial-inspired FSSs in performance enhancement of various antennas is discussed from an application point of view.

5.1 Significance of metamaterial-inspired structures

Metamaterials possessing unusual properties exhibit distinct features [1, 2] that make them suitable for various applications, such as absorbers, cloaking, radar cross section reduction, enhancement of antenna radiation, etc., as mentioned in the introduction [3–6]. Over the years these applications have evolved with the requirement of conformal structures in spacecraft, military aircraft and missiles, radomes, etc [7, 8]. A metamaterial-inspired conformal array structure is shown in figure 5.1. Due to its properties, such as epsilon negative (ENG) and/or mu negative (MNG), metamaterials are preferred over conventional periodic structures in conformal

doi:10.1088/978-0-7503-5422-6ch5

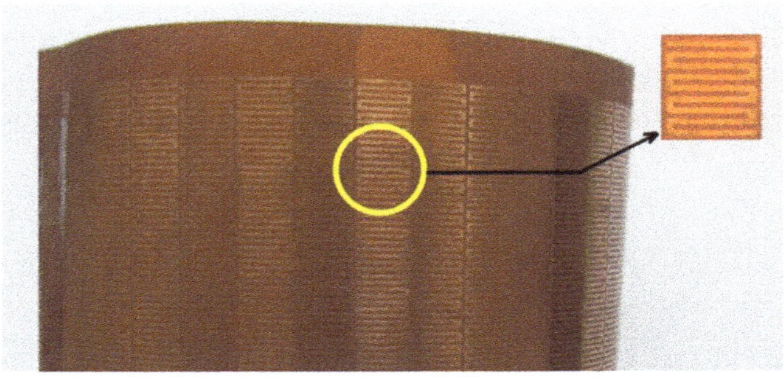

Figure 5.1. Metamaterial-inspired conformal array of frequency selective surface.

applications. A double negative (DNG) metamaterial structure can be realized using a periodic array of thin metallic wires for ENG and split ring resonators for MNG [2].

Conventional periodic structures such as electromagnetic bandgap structures, frequency selective surfaces, and artificial magnetic conductors are composed of resonant unit cells. When a plane wave impinges on metallic patches of the periodic structure, electric current is induced on the elements. The coupling of energy from the plane wave to the surface reaches its maximum value at the resonating frequency where the length of elements is comparable to half-wavelength ($\lambda/2$) or quarter-wavelength ($\lambda/4$) [9]. On the other hand, metamaterials are non-resonant in nature where the unit cell dimension or periodicity is much smaller compared to the wavelength, preferably smaller than $\lambda/4$ or even in some cases $\lambda/10$ or lower, which makes it suitable for curved surfaces as the unit cell response is less affected by the bending [10].

The use of metamaterial-inspired unit cells also leads to smaller inter-element spacing that affects both the operating frequency and bandwidth of the response. A lesser inter-element spacing is preferred as it makes the overall area of the array smaller. A smaller inter-element spacing also gives a wider bandwidth [10, 11]. However, metamaterials have certain disadvantages too, such as:

1. Metamaterials are bulky structures, especially with DNG properties, where both ENG ($\varepsilon < 0$) and MNG ($\mu < 0$) are realized as shown in figure 5.2(a) [12] and figure 5.2(b) [13]. This leads to complex fabrication processes.
2. The plane wave when impinging on a metamaterial structure propagates a significant distance because of its shape, due to which the wave suffers from high dielectric loss.
3. In spite of having smaller inter-element spacing, metamaterial structures possess larger dimensions due to their thickness, which makes it difficult to integrate such structures with other circuits or networks.

In the subsequent sections various methods adopted in designing metamaterial-inspired sub-wavelength FSSs implemented for conformal applications are discussed. The discussion is primarily focused on various kinds of conformal applications. Moreover, integration of metamaterial structures on conventional antennas leading to metamaterial-inspired antennas is also covered in the following sections.

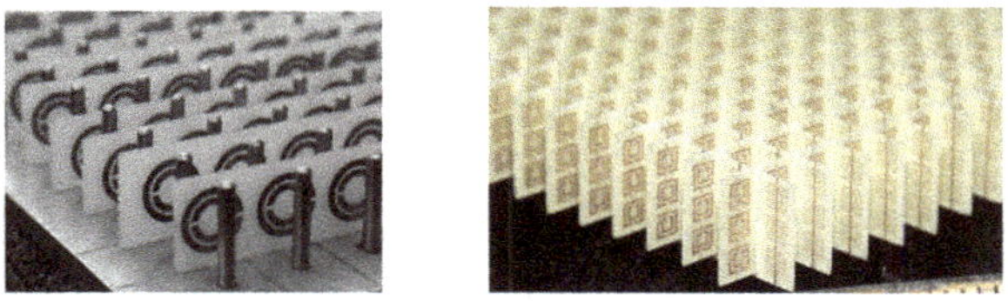

(a) (b)

Figure 5.2. DNG metamaterial made of array of (a) thin metallic wires. Reprinted figure with permission from [12], Copyright (2003) by the American Physical Society. (b) SRRs. Reprinted from [13], with the permission of AIP Publishing.

5.2 Metamaterial-inspired FSSs for conformal applications

In this section, metamaterial-inspired FSS structures developed so far, primarily with a thickness and unit cell dimension much smaller than the wavelength ($\ll\lambda/4$), are discussed for different types of conformal applications. Based on the type of application, such structures can be broadly classified into two categories: conformal FSS-based absorber [16–18] and conformal FSS for non-absorbing applications [36–38] such as polarization conversion, radomes, wearable antennas, etc.

5.2.1 Conformal FSS-based absorber

The traditional microwave absorbers, such as pyramidal absorbers made of low density polyurethane foam as shown in figure 5.3, have a greater thickness [3] and thus are inadequate for integrating with separate structures in most cases. However, metamaterial-inspired absorbers can be realized as periodic arrays on ultrathin substrates [14–16]. Additionally. they can exhibit absorption close to unity over frequency bands including the microwave, terahertz, and optical spectrum. When a metamaterial-inspired FSS is used to realize an absorber on a curved surface, only the centrally oriented unit cells are exposed to normal incidence of the signal, and other unit cells of the FSS are exposed to oblique incidence. Thus angular stability is an essential requirement for the FSS used in designing the conformal absorbers [15].

The use of metamaterial-inspired FSS structures in developing conformal absorbers gained more attention after 2010. Initially conformal absorbers were primarily designed for narrowband applications. Jang *et al* proposed [14] a metamaterial-based absorber with periodic structures of FSS on a cylindrical surface.

$$\mu_{\mathrm{m}}{}^{2}\cos^{2}\theta_{\mathrm{i}} - \mu_{\mathrm{m}}\varepsilon_{\mathrm{m}} + \sin^{2}\theta_{\mathrm{i}} = 0 \tag{5.1}$$

The design method is based on making the reflection coefficient zero at resonating frequency to allow absorption of the signal within the dielectric and conductor. This is done by choosing appropriate values of μ_{m} and ε_{m} in (5.1) for incident angle $\theta_{\mathrm{i}} = 0$ with the combination of split ring and cross shapes in the metamaterial-inspired unit cell as shown in figure 5.4(a). To ensure the highest absorption for other incidence angles on the curved surface, unit cells with different dimensions (R and g) are used at different zones of the cylindrical surface (figure 5.4(b)). The authors have shown 90% maximum absorption rate at 10.85 GHz, as shown in figure 5.4(c).

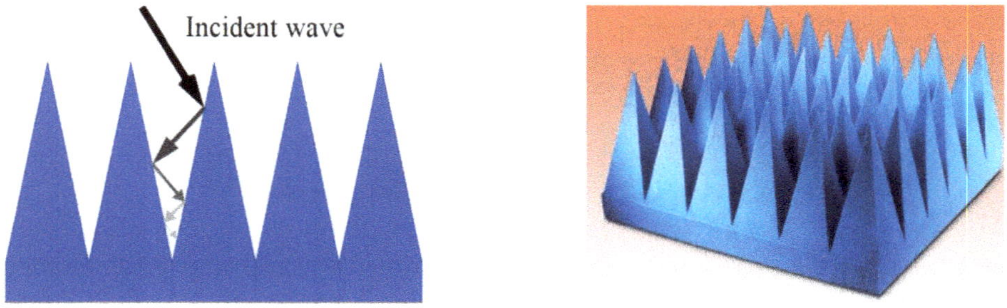

Figure 5.3. Conventional pyramidal RF absorber made of low density polyurethane foam. Adapted from [3].

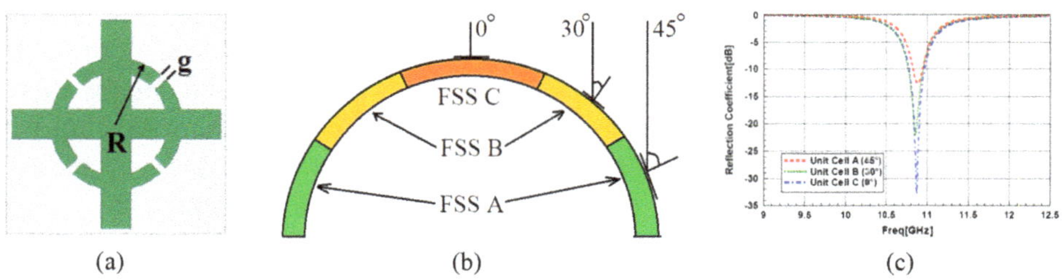

Figure 5.4. (a) Unit cell of the periodic structures (b) Curved surface with the three different metamaterial arrays having different unit cells (c) Reflection coefficient of the unit cells. Adapted with permission from [14].

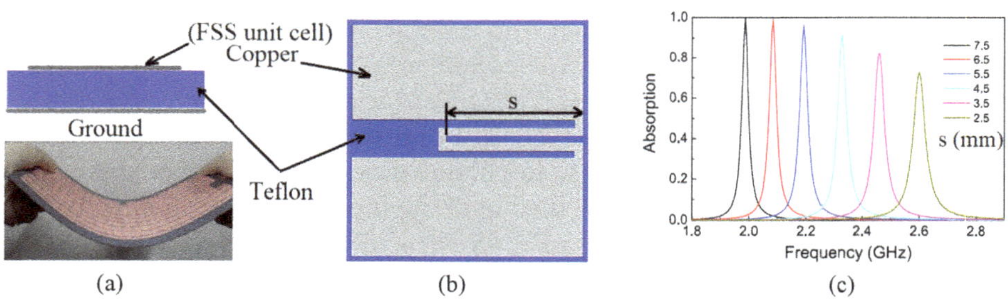

Figure 5.5. (a) Side view of the absorber unit cell and fabricated conformal prototype. (b) Unit cell of the absorber. (c) Absorption coefficient of the metamaterial absorber. Reprinted from [15], with permission of AIP Publishing.

Yoo and his research group proposed [15] a multilayered configuration of metamaterial-inspired FSSs on the top and a ground plane on the other side with a Teflon-based dielectric layer in between. A snake-shaped unit cell was chosen, as shown in figure 5.5(b), with the sub-wavelength dimension on the order of $\lambda_0/12$. The substrate thickness 0.008 λ_0 makes the design ultrathin and suitable for conformal applications, as shown in figure 5.5(a). It was shown by the authors that absorptivity could be increased to 97.8% at 2.07 GHz with larger dimensions of the slot 's'. The absorber was realized under bent conditions and the proposed absorbing periodic structures were found to be flexible as well as elastic.

The absorber as discussed above is, however, sensitive to polarization of the electromagnetic wave impinging on it because of its asymmetric configuration along the vertical axis. Such absorbers are not suitable for various applications, especially in conformal absorbers [16–18], where waves may come from any direction and the electric field might be horizontal or vertical to the unit cell of the absorber array or both components are present [17]. This resulted in the requirement of a polarization-independent absorber with respect to linear polarization. However, microwave absorbers compatible to circular polarization are more suitable for conformal applications in which absorption of the waves impinging on the curved surface at any angle can be ensured. In 2017 Kong *et al* proposed a polarization-insensitive metamaterial-inspired conformal absorbing FSS with wide absorption bandwidth [16]. The unit cell consists of a nested cross square ring structure, leading to three resonances resulting from the rings, as well as the coupling between consecutive ring structures.

The design with the unit cell configuration shown in figure 5.6(a) offers a wide absorption bandwidth as shown in figure 5.6(c). The design is based on ultrathin PI films with $\varepsilon_r = 3.5$ in order to use them for conformal application. Lumped resistors of two different values were incorporated in the unit cell of the FSS in order to achieve a wideband response, ranging between 6.11 and 12.94 GHz, leading to a bandwidth of more than 65% with high rate of absorption up to 90%. The authors have realized the absorber on a conformal surface, as shown in figure 5.6(b), with the use of a 3.5 mm thick layer of rubber that made the design slightly thicker.

Unlike the previously developed metamaterial-inspired absorbers with a narrow-band or wideband response, in 2022 Luo *et al* proposed a multiband absorber based on traditional shapes of unit cells for conformal applications [17]. The unit cell of the top layer consists of a hollowed-out cross inside a circular ring slot (figure 5.7(a)) and the bottom layer is grounded using a metallic screen. The two layers are separated by a flexible polyamide substrate with a permittivity of 3.2 and a loss tangent of 0.009. The design leads to four resonating frequencies at 3.25, 6.89, 9.38 and 10.6 GHz, as shown in figure 5.7(d). The absorber structure has a thickness of 0.436 mm, which is $0.0047\lambda_0$ with respect to the first absorption peak, making it a ultrathin absorber. The metamaterial property of the absorber was investigated and the equivalent permittivity was found to be negative at resonances. The design offers absorption

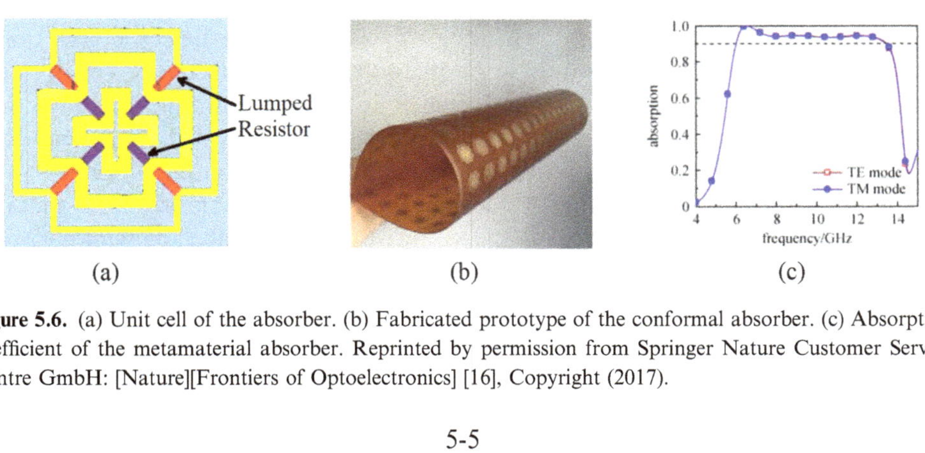

(a) (b) (c)

Figure 5.6. (a) Unit cell of the absorber. (b) Fabricated prototype of the conformal absorber. (c) Absorption coefficient of the metamaterial absorber. Reprinted by permission from Springer Nature Customer Service Centre GmbH: [Nature][Frontiers of Optoelectronics] [16], Copyright (2017).

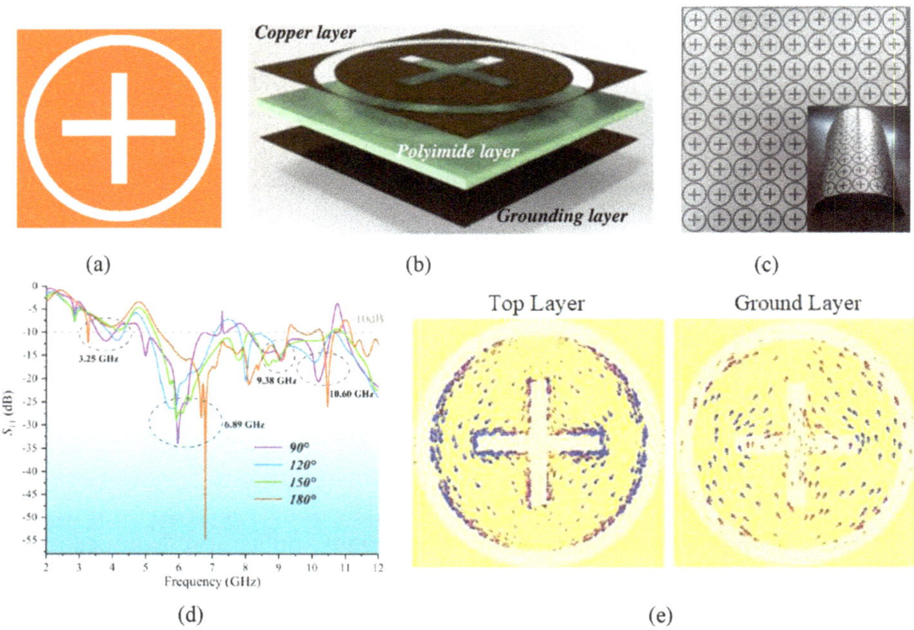

Figure 5.7. (a) Unit cell of the absorber, (b) different layers, (c) fabricated conformal absorber, (d) reflection coefficient of the absorber at various bending angles, and (e) surface current on the layers at 3.25 GHz. Reproduced from [17]

above 90% in the four resonating frequencies. The degree of flexibility of the absorber was tested by the authors at various bending angles of 90°, 120°, 150° and 180° and it was observed that the structure retains it absorptive nature at various bending angles. The conformal fabricated prototype is shown in figure 5.7(c).

In the first three bending angles the first absorption peak exhibits a blue shift and the second absorption peak exhibits a red shift, whereas at 180° bending the calculated absorption peak matches well with the simulated data. The principle of absorption can be better understood from the surface current on the top layer and on the ground plane. As shown in figure 5.7(e), the surface current at 3.25 GHz (lower frequency) flows due to the electric field distribution mainly in the inside and outside of the hollowed-out ring. The electric field distribution in the ground layer is opposite to that in the top layer due to the polarization of the dielectric, and this in turn makes the current at the bottom opposite to the current at the top. The opposite currents at the top layer and ground results in magnetic dipoles leading to a magnetic resonance in the dielectric and consumes the energy of the incident wave.

In recent years the trend of designing conformal FSS-based absorbers with metamaterial properties is primarily focusing on using multiple layers of metallic patches, dielectric substrates, organic glass, resistive sheets, etc., for wider absorption bandwidth. Dai *et al* adopted cascading multiple conductive, dielectric and resistive layers to develop a wideband metasurface-based absorber, as shown in figure 5.8(a), in the range of 8.7 and 38.9 GHz leading to a bandwidth of 127% [18].

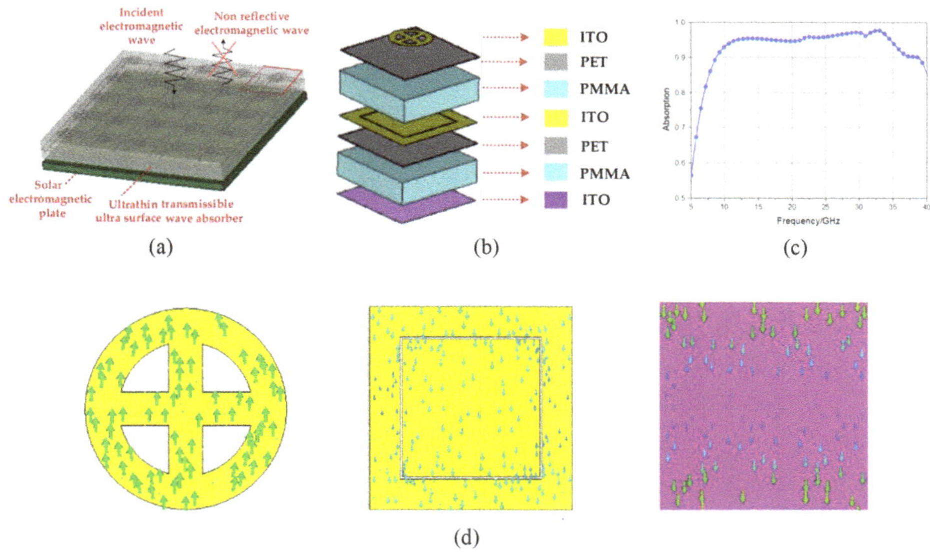

Figure 5.8. (a) Metasurface absorber covered by solar panel, (b) various layers in the absorber, (c) absorption coefficient of the absorber, and (d) surface current distribution on different layers at 22 GHz. Reproduced from [18].

Two layers of ITO resistance film with a square resistance of 5 ohm are placed above 0.1 mm polyethylene terephthalate (PET) layers with the relative permittivity of 3. They are stacked with polymethyl methacrylate (PMMA) plates as shown in figure 5.8(b).

The principle of multi-layer interference cancellation and the corresponding ohmic loss are responsible for the wideband absorption with the absorptivity above 90%. The optically transparent metamaterial absorber is covered with the satellite solar panel. Similar to the previously stated design surface current distribution at various layers of the absorber can reveal the principle of absorption. The opposite surface current flowing on layers 1 and 3 creates the magnetic dipole, which thus leads to absorption of the incident signal.

The use of lumped resistors in developing metamaterial-based absorbers has gained attention in recent years. However, implementation for conformal applications was introduced by Kalraiya *et al* for a wideband response [19]. The proposed unit cell consists of a conventional cross-shaped metallic patch modified with the use of four lumped resistors connecting the four arms to a centrally placed small square patch, as shown in figure 5.9(a). The multilayered structure has a metallic patch on the top of a PET substrate followed by a layer of foam with a ground plane on the bottom. The proposed FSS-based metamaterial absorber has a thickness of $0.088\lambda_0$ and a unit cell with sub-wavelength dimension of $0.14\lambda_0$. Extraction of metamaterial parameters (μ, ε) of the structure shows ENG/MNG characteristics. The design exhibits above 90% absorptivity over a wide range of 3.25–13.6 GHz (figure 5.9(b)), which leads to an ultra wide bandwidth of 122%. The absorber is implemented on a cylindrical surface (figure 5.9(c)) and it is shown by the authors that the proposed design works well for conformal applications with close to 90% absorptivity.

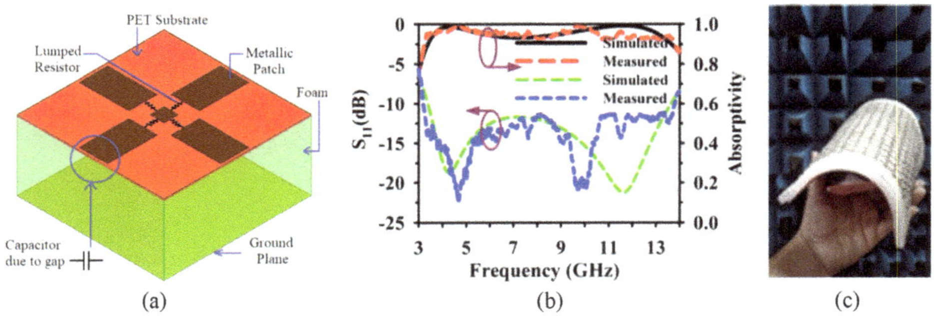

Figure 5.9. (a) Unit cell of the absorber. (b) Characteristics of the absorber. (c) Conformal version of the lumped resistor loaded absorbing structure. Reprinted from [19], Copyright (2022), with permission from Elsevier.

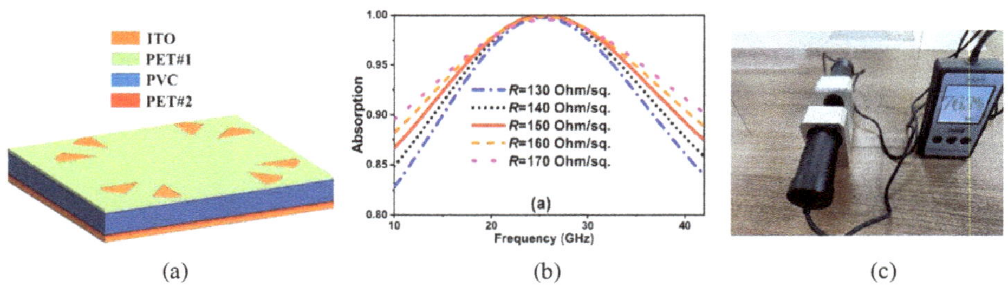

Figure 5.10. (a) Various layers in the metasurface-based absorber. (b) Absorptivity of the absorber for different resistances of ITO layer. (c) Measurement of transparency of the absorber. Reproduced from [20].

Metamaterial-based transparent absorbing structures for ultra broadband conformal applications, including radar cross section (RCS) reduction, were presented by Khan *et al* in 2024 [20]. The absorber is composed of an impedance layer, a substrate layer, and a ground layer. As shown in figure 5.10(a) the resistive ITO layer (150 Ohm Sq^{-1}) is printed on a transparent polyvinyl chloride (PVC) dielectric substrate ($\varepsilon_r = 2.45$). A PET substrate is used as a supporting base for the ITO. The metasurface-based absorber exhibits absorption above 90% in the range of 13.01–39.22 GHz. It was observed that an increase in ITO resistance leads to an increase in 90% absorption bandwidth. The normalized input impedance was found to be close to unity (real part), which justifies the impedance matching between the absorber and air leading to successful absorption of the incident waves. The surface current on the ITO and bottom layer are in opposite directions, resulting in a magnetic dipole responsible for the absorption. The absorber then leads to an RCS reduction of more than 10 dB with a maximum reduction value of 38 dB.

Thus, it can be observed from the studies that the trend of designing metamaterial-inspired FSSs for conformal absorbers started with narrowband structures and continued to multiband, wideband and ultra-wideband designs in recent years. This was accomplished with the cascading of multiple layers of dielectrics, PVCs and resistive films or with the use of lumped resistors. However, in most cases for the conformal implementation of the FSS arrays, cylindrical surfaces are chosen when the application was absorbers.

5.2.2 Conformal FSS for non-absorbing applications

The implementation of metamaterial-inspired FSS structures for conformal applications without absorbing properties are also found in the literature. These include body wearable applications, polarization converters, filters for antennas, radomes, etc [21–23]. The structures can be reconfigurable in some cases. However, in most cases ultrathin structures have been used.

The implementation of a metamaterial-inspired bandpass FSS for conformal applications was proposed by Haghzadeh and Akyurtlu in 2016 [21]. A square loop-shaped patch and metallic grids are used on the two layers, respectively, as shown in figure 5.11(a). It was demonstrated that with the use of an interdigitated capacitor (IDC) and by printing the space between fingers of the IDCs with different dielectrics, the transmission pole was tuned leading to miniaturization of the FSS (figure 5.11(b)). Barium strontium titanate (BST) or cyclic olefin copolymer (COC) is used as the dielectric here. Initially all the IDCs were printed with the dielectric for a single frequency band (figure 5.11(c)). Later on with irregular printing, every alternative IDCs dual-band response was achieved without any change in the unit cell shape or dimension. A unit cell dimension of $0.1\lambda \times 0.1\lambda$ was chosen, leading to sub-wavelength structure. Further tuning up to 21% was achieved with the dielectric variation. The FSS was implemented on a curved structure (figure 5.11(c)) and a satisfactory response was achieved, however, with increased insertion loss.

A novel approach of designing flexible as well as stretchable meta-skin based on meta-atoms followed by metamaterials can be found in the literature by Yang *et al* [22]. Initially the meta-skin is designed using split ring resonator based meta-atoms with EGaIn as the metal and silicone ecoflex as the substrate.

The unit cell can be seen in figure 5.12(a). With the H field applied normal to the ring, it exhibits magnetic resonance. Six such meta-skins are cascaded (figure 5.12(b))

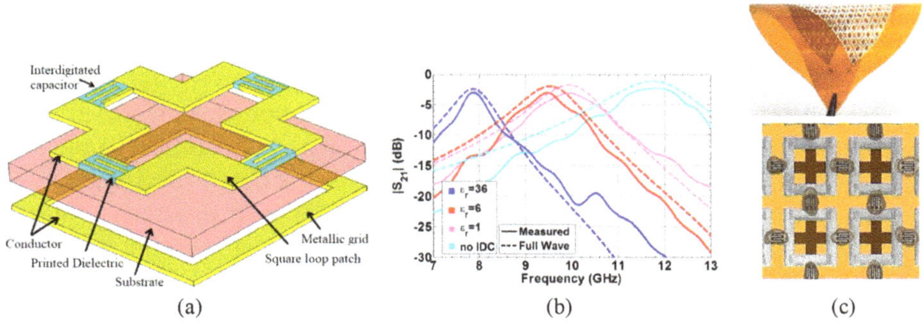

Figure 5.11. (a) Unit cell of the loop grid FSS with interdigitated capacitor (IDCs). (b) Transmission response of the FSS. (c) FSS with BST/polymer filled IDCs and conformal prototype. Reprinted from [21], with permission of AIP Publishing.

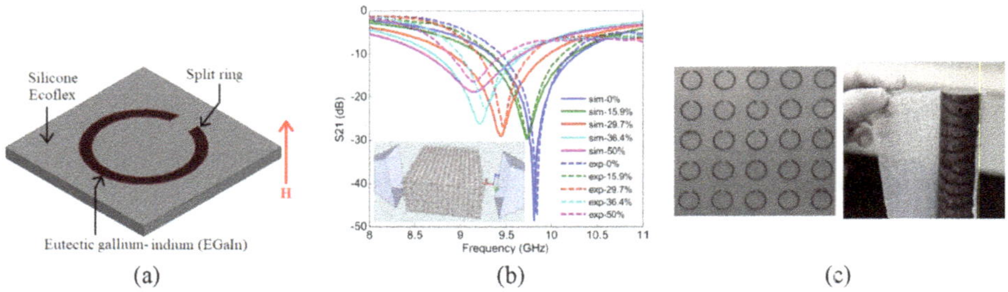

Figure 5.12. (a) Meta-skin unit cell. (b) Transmission response of the metamaterial-based meta-skin for various stretching conditions. (c) Dielectric nylon rod wrapped by the meta-skin. Adapted from [22]. CC BY 4.0.

with a spacing of 3 mm between each and when excited using horn antenna, exhibits resonance at 9.84 GHz. The authors have demonstrated and tested the conformal prototype by wrapping a nylon rod with the meta-skin (figure 5.12(c)). The meta-skin was stretched and it was shown how the resonance was shifting. The design is suitable for wearable applications in medical telemetry due to its highly flexible and stretchable properties.

Another significant application of metamaterial-based bandpass FSS structures is found in conversion of polarization of the RF signals [23–25]. Microwave or mm-wave signals emitted from the antenna are often desired to be modified in terms of horizontal to vertical polarization or vice versa [23]. Additionally, certain polarizers are desired to transform linear polarization into circular polarization (LHCP or RHCP) [24]. Although FSS-based polarization converters are primarily designed on planar surfaces, for applications involving curved structures such as aircraft, radomes, etc., they should be implemented on a conformal surface. For this purpose, polymer- or RT Duroid-based substrates are frequently used because of their higher degree of flexibility. However, in order to increase the mechanical strength of the FSS screens during bending, further substrate-integrated waveguide (SIW) technology is implemented in the designs. This in turn increases the Q factor of the bandpass response, leading to higher roll-off factor and selectivity.

The application of metamaterial-inspired flexible FSS structures for polarization conversion was proposed by Varikuntla and Singaravelu in 2018 [25]. Y-shaped slots of sub-wavelength dimension are loaded on both top and bottom sides of the dielectric substrate in the design, which gives the bandpass response as shown in figure 5.13. A novel approach in the design is implementation of substrate-integrated waveguide (SIW) technology in the unit cell, which adds an advantage of enhancing the selectivity in the transmission response, as can be clearly observed in figure 5.13(b). In the dual-layer design the authors have shown that by using differently rotated Y slots on the two layers, a single- or dual-band response can be achieved or polarization conversion by $+90°$ or $-90°$ can be obtained. The reflectance (r_{xx}) and transmittance (t_{xy}) of the design shows a sharp roll-off unlike the previous designs (figure 5.13(b)). A flexible Rogers RT Duroid 5880 substrate is chosen for the conformal application (figure 5.13(c)). For this purpose a smaller thickness of the substrate was chosen.

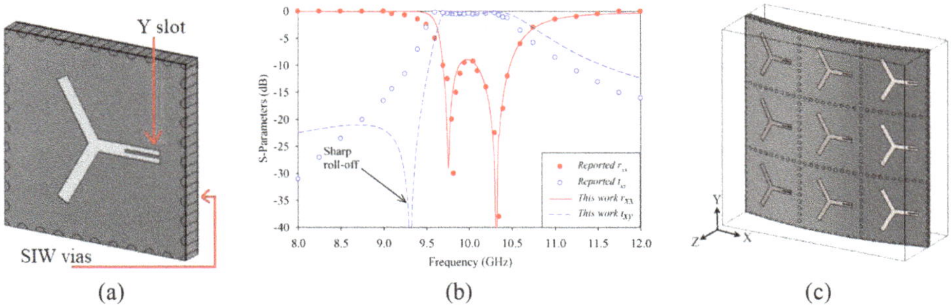

Figure 5.13. (a) Meta-skin unit cell. (b) Transmission response of the metamaterial-based meta-skin for various stretching conditions. (c) Dielectric nylon rod wrapped by the meta-skin. [25] John Wiley & Sons. [© The Institution of Engineering and Technology.]

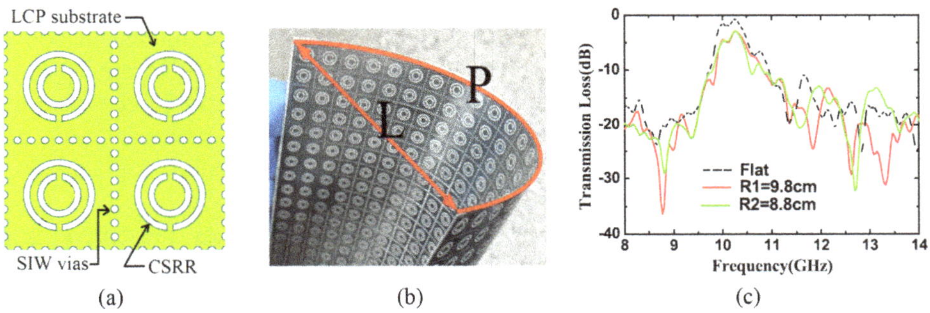

Figure 5.14. (a) CSRR-based SIW loaded unit cell of the FSS. (b) Fabricated prototype of the FSS on polymer substrate. (c) Comparison of the response for flat and conformal FSS. Adapted from [26]. CC BY 4.0.

Another design of FSS-based conformal polarizer with the application of complementary split ring resonators (CSRR) and SIW was proposed by Li *et al* recently [26]. The unit cell of the FSS as shown in figure 5.14(a) consists of CSRR-based slots placed on both sides of a liquid crystal polymer (LCP) substrate of 50 μm thickness. CSRR is frequently used in metamaterial-inspired structures. LCP, being a polymer, is suitable in fabricating designs over a curved surface. The authors have implemented substrate-integrated waveguide technology in the design, which led to a high Q factor and thus better selectivity in the transmission response. An extinction ratio of 18.3 dB was achieved between TE and TM polarization incidence. The design shows angular stability up to 60°, making it suitable for conformal applications. The design was fabricated and bending performance was investigated for different radii. The bending radius (R) for the flexible FSS was calculated from the curvature length (P) and the end to end distance (L) after bending using the following relation:

$$\frac{L}{2R} = \sin\left(\frac{P}{R}\right)$$

The insertion loss of the FSS under bending conditions got elevated to 3 dB from 0.75 dB under flat conditions. However, the transmission response of the curved structure as shown in figure 5.14(c) proves its suitability for conformal applications.

5.3 Metamaterial-inspired antennas for conformal applications

The advent of modern technologies in wireless communication such as Internet of Things (IoT), 5G network, Wi-Fi 6, etc, has raised the demand for higher data rates, system reliability, wide bandwidth, and higher efficiency [27]. Antennas used for such technologies need to satisfy all necessary requirements and that too with low profile. For reliable output and increased efficiency the antennas should also have higher gain and controlled beamwidth. Such requirements can be fulfilled with the modification of the antenna design using various metamaterial-inspired structures such as electromagnetic bandgap structures (EBGs), frequency selective surfaces (FSSs), artificial magnetic conductor (AMCs), metasurfaces, negative refractive index (NRI) metamaterials, etc [28–31]. Such structures can be loaded in conventional antennas either on the same plane or at spacing from the substrate leading to metamaterial-inspired antennas. Many such works are available in the literature. However, applications involving wearable antennas, aircraft, radomes, etc., need curved structures. This section primarily focuses on conformal applications of various kinds of metamaterial-inspired antennas.

The left-handed structure showing metamaterial properties presented in the preceding section, which is designed using metallic wires and split rings, exhibits loss and limited bandwidth. As a solution to this problem a modified artificial left-handed structure was proposed [28, 32] termed a composite right/left-handed transmission line (CRLH). CRLH transmission lines can be designed by modifying the equivalent circuit of a left-handed structure followed by implementation of the same in the structure.

The LC network (figure 5.15(a)) in the circuit consists of series capacitance C_L and shunt inductance L_L that are required to provide the left-handed characteristics [32]. The series inductance L_R is included to account for the current flow in the metallization and the shunt capacitance C_R accounts for the potential difference developed between the signal traces and the ground plane. Dispersion response of the CRLH structure exhibits propagation constant $\beta < 0$ at lower frequencies and is thus left-handed, whereas at frequencies exceeding the stop-bandgap it shows right-

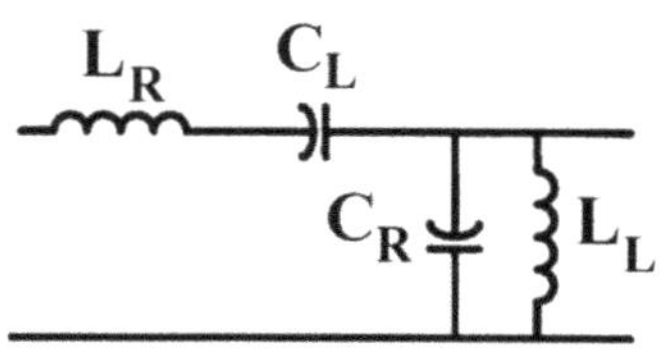

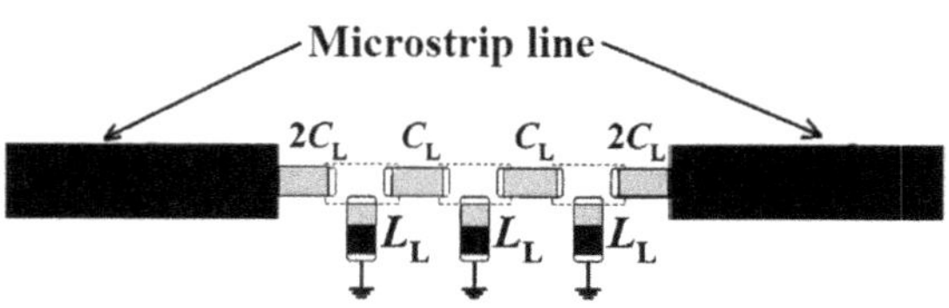

Figure 5.15. (a) Equivalent circuit model of the CRLH structure. (b) Realization of CRLH transmission line with chip capacitors and inductors in microstrip line. [32] John Wiley & Sons. [Copyright © 2006 John Wiley & Sons, Inc.]

handed properties where β becomes zero [28]. In reality CRLH transmission lines (CRLH-TL) can be realized in various ways; one of them is using a microstrip line with the inclusion of chip inductors and capacitors as shown in figure 5.15(b). Due to the low profile of the components, the length of the CRLH-TL can be reduced by simply controlling the length of the microstrip line. CRLH transmission lines have been used as a metamaterial-inspired structure along with the antennas for miniaturization by many researchers [28–31].

Yan *et al* proposed [30] the application of CRLH structure-based metamaterials for miniaturization of dimensions of a dual-band wearable antenna operating at 2.4 and 5.2 GHz in 2014. At the centre of the rectangular loop-shaped antenna patch, two mushroom-like structures are introduced in order to form the CRLH-TL structure (figure 5.16(a)). A 6 mm thick felt substrate is used for the antenna, which makes the prototype flexible (figure 5.16(b)) and suitable for wearable applications. At 2.4 GHz, the antenna exhibits a significant miniaturization in the dimension down to $\lambda_0/6 \times \lambda_0/6 \times \lambda_0/20$. The prototype is tested under bending conditions and was found suitable for wearable applications in the ISM bands as depicted in figure 5.16(c). Under bending the conformal antenna shows no significant changes in the characteristics, including an SAR value less than 0.7 W kg^{-1} averaged over 10 g of tissue, which is far below the regulated European threshold of 2 W kg^{-1}.

In the above mentioned design, a comparatively thicker substrate was used to incorporate the vias needed to realize inductors in the CRLH lines. This problem was later overcome by Sahoo and Vakula [31] with the use of meander lines to realize the CRLH structure in the plane of the dual-band antenna. The design uses three such cells with spacing placed inside the radiating patch as shown in figure 5.17 (a). Moreover, the bandwidth was enhanced with the use of a fractal inductor between ground and patch. Later on, two parasitic meander lines were introduced above the patch leading to a dual-band response at 2.4 and 3.5 GHz as shown in figure 5.17(b). The CRLH structures used here are of sub-wavelength dimension ($0.06\lambda_g$). An impedance bandwidth up to 50% was achieved for the first band. The design was placed above a cylindrical structure with 15 mm radius for conformal application and the antenna characteristics were almost the same under bending conditions.

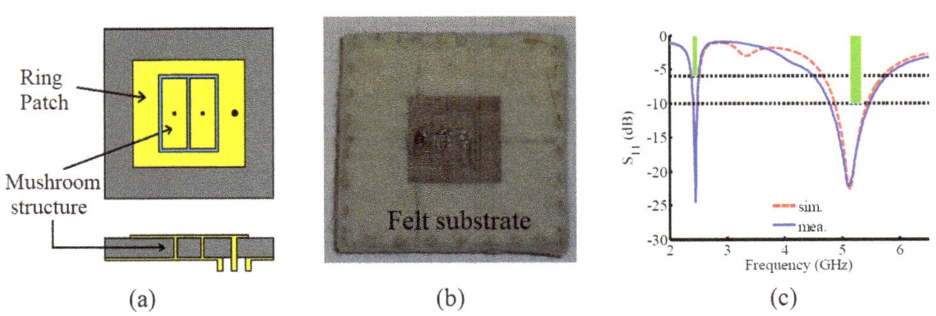

Figure 5.16. (a) CRLH loaded antenna with mushroom structure. (b) Antenna fabricated on flexible felt substrate. (c) Simulated and measured reflection coefficient of the antenna. Reproduced from [30]. CC BY 4.0.

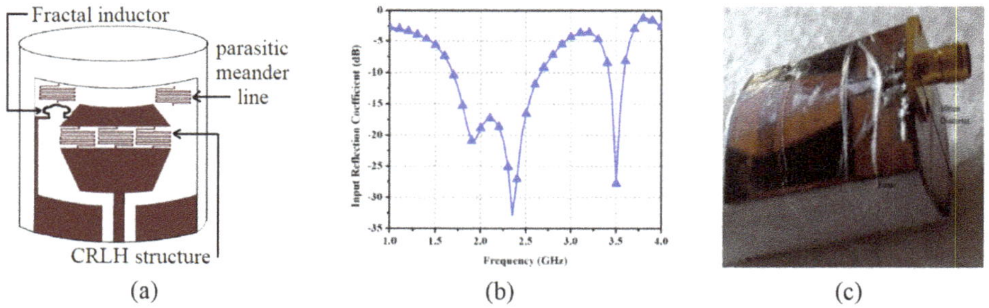

Figure 5.17. (a) Conformal antenna with CRLH structure and parasitic elements. (b) Reflection coefficient of the metamaterial-inspired antenna. (c) Fabricated conformal antenna. [31] John Wiley & Sons. [© 2019 The Institution of Engineering and Technology]

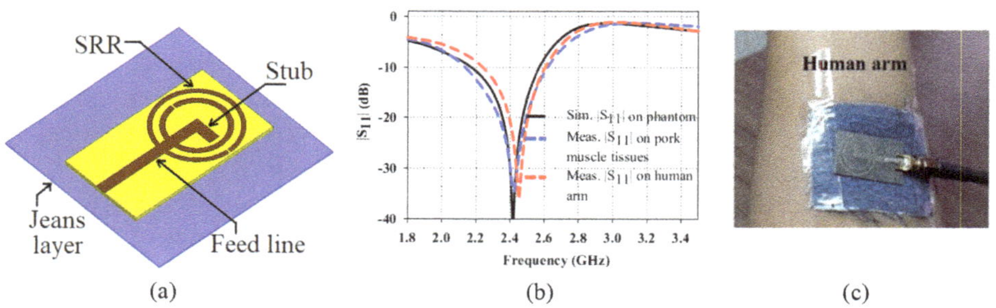

Figure 5.18. (a) Antenna loaded with SRR and stub along with jeans cloth below. (b) Reflection coefficients of the metamaterial-inspired antenna. (c) Fabricated antenna on human arm. Reprinted by permission from Springer Nature Customer Service Centre GmbH: [Nature][Wireless Personal Communications] [33], Copyright (2019).

Another approach of realizing metamaterial-inspired antennas is the incorporation of split ring resonators (SRRs) in the antenna radiating plane or below [33–35]. Split ring resonators are used in designing metamaterial structures, as was discussed in section 4.1, in order to exhibit negative permeability (MNG). Eventually SRRs were used by researchers also in conformal antennas to achieve miniaturization of the dimension, circular polarization, etc.

Chaturvedi *et al* proposed [33] a SRR-loaded monopole antenna at 2.45 GHz for wireless body area network (WBAN) applications (figure 5.18). The SRR has a subwavelength dimension on the order of $\lambda/7$. The ground plane is placed on the back side with tapering for better impedance matching. Apart from SRR for the inclusion of an L-shaped stub in the monopole reduced the antenna dimension by 7.5%. Metamaterial properties of the SRR with negative permeability are presented in the work. A thin layer of jeans cloth (thickness = 1 mm) is placed below the antenna in order to achieve further miniaturization as well as biocompatible isolation from the body when placed above skin. The authors have tested the antenna both on pork muscle tissue and human arm (figure 5.18(c)) showing satisfactory performance of the antenna for conformal applications.

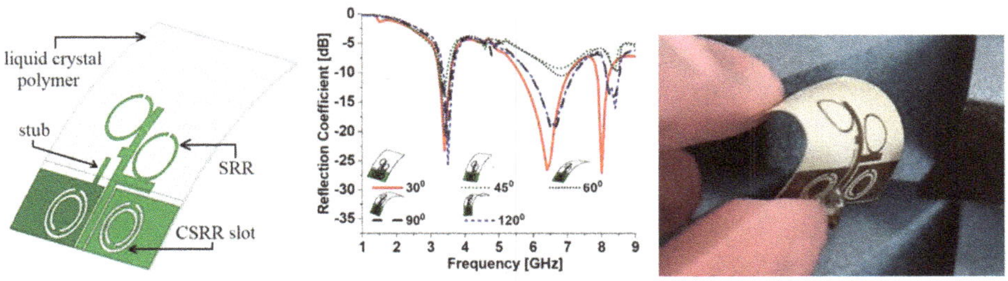

(a) (b) (c)

Figure 5.19. (a) Antenna with SRR-loaded stub along with CSRR in ground. (b) Reflection coefficients of the antenna under bending conditions. (c) Fabricated conformal antenna. [34] John Wiley & Sons. [© 2019 Wiley Periodicals, Inc.].

A different kind of application of SRR-based metamaterial-inspired conformal antennas was proposed by Rao *et al* recently, where circular polarization is achieved in a CPW fed conformal antenna at 3.5 and 5.8 GHz [34]. The antenna as shown in figure 5.19(a) consists of a simple monopole with CPW and the monopole is loaded with stubs. The stubs are further connected to two SRRs that are placed asymmetrically. Moreover, complementary split ring resonator (CSRR)-shaped slots are included in the ground plane on both sides. The asymmetrically placed SRR and CSRR slots lead to circular polarization. This is confirmed by the authors with the plot of axial ratio below 3 dB and current on the antenna at four different phase angles. Performance of the antenna is verified for different bending angles of 30°, 45°, 60° and the results are close to the planar one as shown in figure 5.19(b). The antenna is fabricated (figure 5.19(c)) on liquid crystal polymer (LCP) substrate in order to achieve a high degree of flexibility in the conformal design.

References

[1] Cui T J, Alù A and Pendry J B 2023 Guest editorial special issue on metamaterials, metadevices, and applications *IEEE Trans. Microw. Theory Tech.* **71** 3229–34

[2] Marqués R, Martin F and Sorolla M 2011 *Metamaterials with Negative Parameters: Theory, Design, and Microwave Applications* (New York: Wiley)

[3] Acquaticci F, Yommi M M, Gwirc S N and Lew S E 2017 Rapid prototyping of pyramidal structured absorbers for ultrasound *Open J. Acoustics* **7** 83

[4] Pallavi M, Kumar P, Ali T and Shenoy S B 2022 Modeling of a negative refractive index metamaterial unit-cell and array for aircraft surveillance applications *IEEE Access* **10** 99790–812

[5] Amiri M, Tofigh F, Shariati N, Lipman J and Abolhasan M 2020 Review on metamaterial perfect absorbers and their applications to IoT *IEEE Internet Things* J **8** 4105–31

[6] Fan Y, Wang J, Fu X, Li Y, Pang Y, Zheng L, Yan M, Zhang J and Qu S 2019 Recent developments of metamaterials/metasurfaces for RCS reduction *EPJ Appl. Metamater.* **6** 15

[7] Tadesse A D, Acharya O P and Sahu S 2020 Application of metamaterials for performance enhancement of planar antennas: a review *Int. J. RF Microw Comput.-Aided Eng.* **30** e22154

[8] Krishnamoorthy K and Narasimhadhan A V 2020 A high gain zero index metamaterial for radome applications *2020 IEEE Int. Conf. on Electronics, Computing and Communication Technologies (CONECCT)* pp 1–5 (Piscataway, NJ: IEEE)
[9] Munk B A 2005 *Frequency Selective Surfaces: Theory and Design* (New York: Wiley)
[10] Anwar R S, Mao L and Ning H 2018 Frequency selective surfaces: a review *Appl. Sci.* **8** 1689
[11] Bayatpur F 2009 Metamaterial-inspired frequency-selective surfaces (Doctoral Dissertation, University of Michigan)
[12] Whitesides G M 2002 Organic materials science *MRS Bull.* **27** 56–65
[13] Wilson J D and Schwartz Z D 2005 Multifocal flat lens with left-handed metamaterial *Appl. Phys. Lett.* **86** 021113
[14] Jang Y, Yoo M and Lim S 2013 Conformal metamaterial absorber for curved surface *Opt. Express* **21** 24163–70
[15] Yoo Y J, Zheng H Y, Kim Y J, Rhee J Y, Kang J H, Kim K W, Cheong H, Kim Y H and Lee Y P 2014 Flexible and elastic metamaterial absorber for low frequency, based on small-size unit cell *Appl. Phys. Lett.* **105** 041902
[16] Kong X, Xu J, Mo J J and Liu S 2017 Broadband and conformal metamaterial absorber *Front. Optoelectron* **10** 124–31
[17] Luo Z, Ji S, Ji Z, Liu Z and Dai H 2022 An ultra-thin flexible conformal four-band metamaterial absorber applied in S-/C-/X-band *Phys. Scr.* **97** 045813
[18] Dai H, Li S, Dong P and Ma Y 2023 Design of an ultra-wideband transparent wave absorber *Materials* **16** 5962
[19] Kalraiya S, Chaudhary R K and Abdalla M A 2022 Resistor loaded wideband conformal metamaterial absorber for curved surfaces application *AEU-Int. J. Electron. Commun.* **143** 154033
[20] Khan H A, Majeed A, Zahra H *et al* 2024 Transparent conformal metasurface absorber for ultrawideband radar cross section reduction *J. Phys.* D **57** 135105
[21] Haghzadeh M and Akyurtlu A 2016 All-printed, flexible, reconfigurable frequency selective surfaces *J. Appl. Phys.* **120** 184901
[22] Yang S, Liu P, Yang M, Wang Q, Song J and Dong L 2016 From flexible and stretchable meta-atom to metamaterial: a wearable microwave meta-skin with tunable frequency selective and cloaking effects *Sci. Rep.* **6** 21921
[23] Liu D J, Xiao Z Y, Ma X L and Wang Z H 2015 Broadband asymmetric transmission and multi-band 90° polarization rotator of linearly polarized wave based on multi-layered metamaterial *Opt. Commun.* **354** 272–6
[24] Mutlu M, Akosman A E, Serebryannikov A E and Ozbay E 2011 Asymmetric chiral metamaterial circular polarizer based on four U-shaped split ring resonators *Opt. Lett.* **36** 1653–5
[25] Varikuntla K K and Singaravelu R 2018 Ultrathin design and implementation of planar and conformal polarisation rotating frequency selective surface based on SIW technology *IET Microw. Antennas Propag.* **12** 1939–47
[26] Li W, Lan Y, Wang H and Xu Y 2021 Microwave polarizer based on complementary split ring resonators frequency-selective surface for conformal application *IEEE Access* **9** 111383–9
[27] Arya K V, Bhadoria R S and Chaudhari N S 2018 *Emerging Wireless Communication and Network Technologies* (Berlin: Springer)

[28] Caloz C 2006 Dual composite right/left-handed (D-CRLH) transmission line metamaterial *IEEE Microw. Wirel. Compon. Lett.* **16** 585–7

[29] Alibakhshikenari M *et al* 2020 A comprehensive survey of 'metamaterial transmission-line based antennas: design, challenges, and applications *IEEE Access* **8** 144778–808

[30] Kai Z, Soh Ping J and Sen Y 2020 Meta-wearable antennas—a review of metamaterial based antennas in wireless body area networks *Materials* **14** 149

[31] Sahoo R and Vakula D 2019 Compact metamaterial inspired conformal dual-band antenna loaded with meander lines and fractal shaped inductor for Wi-Fi and WiMAX applications *IET Microw. Antennas Propag.* **13** 2349–59

[32] Caloz C and Itoh T 2005 *Electromagnetic Metamaterials: Transmission Line Theory and Microwave Applications* (New York: Wiley)

[33] Chaturvedi D and Raghavan S 2019 A compact metamaterial-inspired antenna for WBAN application *Wirel. Pers. Commun.* **105** 1449–60

[34] Venkateswara Rao M, Madhav B T, Anilkumar T and Prudhvinadh B 2020 Circularly polarized flexible antenna on liquid crystal polymer substrate material with metamaterial loading *Microw. Opt. Technol. Lett.* **62** 866–74

[35] Devarapalli A B and Moyra T 2023 Design of a metamaterial loaded W-shaped patch antenna with FSS for improved bandwidth and gain *Silicon* **15** 2011–24

[36] Nisanci M H and de Paulis F 2021 Easy-to-design-and-manufacture frequency selective surfaces for conformal applications *IEEE Antennas Wirel. Propag. Lett.* **20** 753–7

[37] Chatterjee A and Parui S K 2018 Beamwidth control of omnidirectional antenna using conformal frequency selective surface of different curvatures *IEEE Trans. Antennas Propag.* **66** 3225–30

[38] Bilal M, Saleem R, Abbasi Q H, Kasi B and Shafique M F 2020 Miniaturized and flexible FSS-based EM shields for conformal applications *IEEE Trans. Electromagn. Compat.* **62** 1703–10

Ayan Chatterjee, Snehasish Saha, Sushanta Sarkar and Partha Pratim Sarkar

Chapter 6

Metamaterial-based reconfigurable antenna and FSS design

6.1 Introduction

In this chapter, the approaches to design the reconfigurable antennas and FSSs using metamaterials and their application is being discussed. One can obtain reconfigurability (active tunability and switching) of the electromagnetic properties of metamaterials by modifying the size, composition, and form of individual meta-atom or metamolecule resonators. The near-field interactions are altered concurrently for this modification of unit cell structure. This reconfigurable property can also be attained either by shifting arrays of metamolecules to produce a three-dimensional metamaterial lattice or by altering the relative positions of the rows in the meta-molecular structure [1, 2]. Micro-electro-mechanical systems (MEMS) were first employed to tune transmission lines for electromagnetic metamaterials [3, 4]. After this phase, MEMS actuators were introduced to make the metamolecules reconfigurable as well as tunable negative refractive index metamaterial arrays by adjusting each component's resonating properties [5]. The metamaterial structure with reconfigurable properties [6, 7] at the THz range were originally produced by fabricating two-dimensional arrays of split-ring resonators on bi-material cantilevers designed to flex out of plane in response to a thermal stimulus. This type of thermally activated structures may be used in THz and infrared detectors. Among the most intricate designs with MEMS actuators, it is possible to make the THz metamaterial dynamically adjusted by fabricating the array of metamaterial on MEMS-driven silicon-on-insulator. The strength of the dipole–dipole coupling can be continuously adjusted by varying the distance between the two rings using the MEMS actuators, which enables effective electromagnetic response customization. This anisotropic metamaterial's polarization eigenstates can also be switched thanks to the reorganization of metamolecules [1]. Plasmonic resonator arrays on

doi:10.1088/978-0-7503-5422-6ch6

stretchable, flexible polymer substrates provide a useful means of dynamically tuning the response of photonic metamaterials and enable their fabrication on curved shapes.

By introducing a conductive liquid, such as mercury, through the network of metamolecules, microfluidics can also be used for reconfiguring microwave meta-devices by altering their EM characteristics (figure 6.1(a)) [8]. Moreover, an electro-statically driven photonic metamaterial comprising a gold plasmonic nanowire pattern deposited on a dielectric membrane has also been explored (figure 6.1(b)) [9]. Operating within the wavelength range used for optical telecommunication, it can be utilized as a MHz-bandwidth modulator with minimal power consumption and perform nonvolatile switching to provide high-contrast transmission changes. That's why, in the future, MEMS, NEMS, and micro-/nano-fluidics technologies will significantly influence the metamaterial industry. As nanoscale components may have mechanical oscillation frequencies in the GHz range, high bandwidth driving

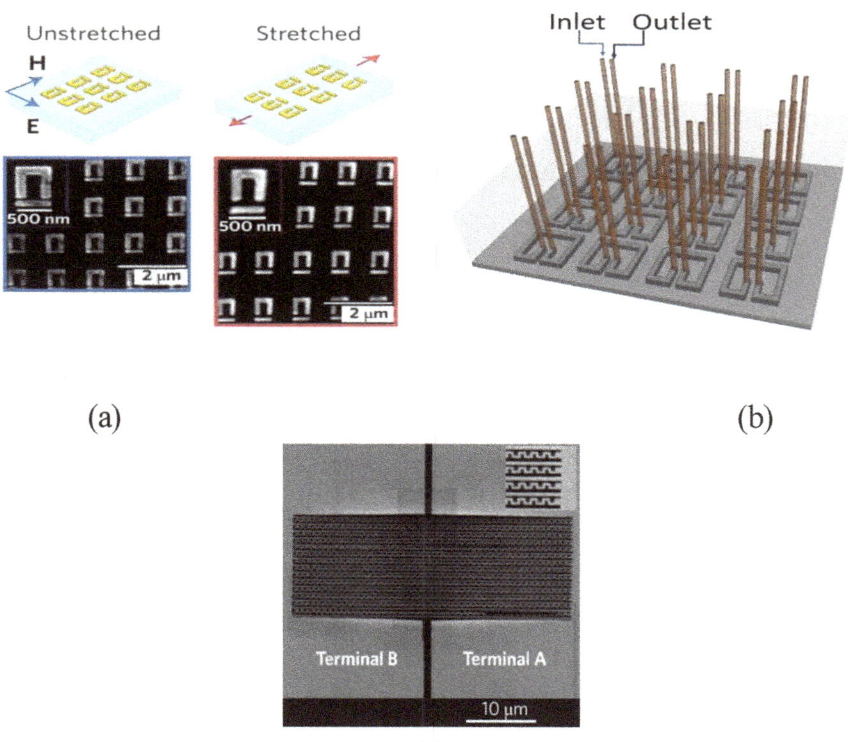

(c)

Figure 6.1. (a) Through substrate stretching, resonator component manipulation allows for the tuning of photonic metamaterials produced on polymer films. The incident electric field (magnetic field) is indicated by the letters E and H. (b) Diagram of a split-ring resonator-shaped microwave metamaterial that can be switched by introducing liquid metallic mercury into the capillary array. (c) Electrostatic driving of plasmonic metamaterials produced on dielectric threads cut from a silicon nitride membrane to megahertz frequencies is possible. An SEM inset at the terminal end displays a portion of the array. Reprinted by permission from Springer Nature Customer Service Centre GmbH: [Nature][Nature Materials] [1], Copyright (2012).

of appropriately structured metamolecule arrays based on subwavelength-sized cantilevers is possible. This method can directly compete with electro-optical modulators in certain applications because it allows for high-density integration while also providing low-voltage operation. The production of plasmonic resonator arrays on elastic, flexible polymer substrates provide a useful method for dynamically adjusting the response of photonic metamaterials and enables their fabrication on curved forms. Moreover, a photonic metamaterial driven by electrostatic forces has been developed, consisting of a gold plasmonic nanowire pattern fabricated on a dielectric membrane (figure 6.1(c)) [10]. It can be utilized as a megahertz-bandwidth modulator, consuming only a few microwatts of power, and can perform nonvolatile switching, giving high-contrast transmission change. It operates in the optical telecommunication range of wavelengths. Therefore, MEMS, NEMS, and micro-/nano-fluidics technologies will have considerable impact on meta-devices in the future. The mechanical oscillation frequencies of nanoscale components could be in the gigahertz range, allowing properly engineered arrays of metamolecules based on subwavelength-sized cantilevers to be driven at high bandwidth. This method can directly compete with electro-optical modulators in some applications since it allows for high-density integration while also providing low-voltage operations.

6.2 Some tuning techniques for getting reconfigurability

Through the use of resonant phenomena, metamaterials and meta-surfaces are engineered to exhibit a variety of exotic properties. Due to this, metamaterials typically exhibit dispersive electromagnetic responses and only exhibit desired properties at extremely narrow-band frequency ranges. As a result of this, it has been known since the beginning of metamaterial research that, in order to create metamaterials that are practically useful, one must be adept at spontaneously adjusting the behavior of the composite structures. Alternatively, when the broad-band performance is not important, one may want to change the properties of materials at a given frequency so that the electromagnetic waves can be dynamically manipulated. This can be used, for example, in switching and modulation applications, making the ability to tune material properties highly desirable. There is a substantial amount of work done on tunable metamaterials to date, with several reviews published [1, 11–14] and different journals having made this a topic of special issues [15].

First, we would like to define what we call tunable metamaterials [16–18]. Some authors say that they 'tune' the properties of materials by fabricating a new sample with different parameters, or by changing the operating frequency, or by changing the incidence angle of the electromagnetic waves. Clearly, in these examples the properties of each given material are not changed, with either a totally new material fabricated or by changing the excitation conditions. Here, we will call metamaterials tunable if their properties can be changed by external influence, e.g., by control voltage, or by temperature, or by magnetic field, etc. Overall, we can distinguish three distinct tuning mechanisms that can be applied for changing the material properties:

- **Tuning by modifying the structural geometry:** Achieved by mechanically deforming the shape of the constituent elements or their mutual arrangement in the metamaterial, which affects the overall properties due to mutual coupling between these elements [10], e.g., elastic deformation of the structure.
- **Tuning by modifying the constituent materials:** Achieved by changing the properties of the materials composing the individual meta-atoms, e.g., actively changing the conductivity of semiconductors by injecting free electrons into them.
- **Tuning by adjusting of the surrounding environment:** Achieved by immersing the metamaterial in an environment, which properties can change, e.g., by a liquid crystal (LC). Each of these tuning mechanisms can be realized in a different way, which is also specific to the frequency range where this metamaterial operates.

  Each of these tuning mechanisms can be realized in a different way, which is also specific to the frequency range where this metamaterial operates.

6.3 Tunable microwave metamaterials

There is a substantial amount of work done on tunable metamaterials and composite right and left-handed transmission lines in the microwave frequency range, and we can see three main reasons for this. Firstly, we have a large number of materials and commercially available components whose properties change under external influence. Secondly, the fabrication of metamaterials is much easier at microwave frequencies, where constituent components are quite large. Thirdly, immediate applications of metamaterials seem to be more feasible in this frequency range with several products currently being developed. The most obvious way to tune metamaterials or transmission lines in microwaves is to use semiconductor components, such as diodes. This was demonstrated, e.g., in references [19–21]. The tunability obtained by this approach is quite large, since the parameters of, e.g., varactor diodes can be changed by almost an order of magnitude. An extension of this approach is based on the use of photodiodes for biasing varactor diodes. When such structures are illuminated by light, the photodiodes generate voltage, and the microwave response of the metamaterial changes. This may be convenient for creating light-controlled tunable metasurfaces, as it was demonstrated in reference [22]. Moreover, such tunability not only controls linear properties of the metamaterial, but also its nonlinear properties [23].

Another way of making tunable metamaterials at low frequencies is by utilizing superconductors. As an example, the properties of superconductor metamaterials in reference [24] were controlled by three different means in order to make tunable metamaterials, where authors used temperature, dc magnetic field, and RF magnetic field to change the properties of metamaterials. Superconductors in metamaterials can further be used in Josephson junction configuration, and this opens the path for the applications of metamaterials for solving quantum problems. As an example, in reference [25], the authors used metamaterial-type superconducting quantum interference devices (SQUIDs) in order to achieve a tunable amplifier that can be used for amplifying quantum signals without adding noise.

Furthermore, tunable metamaterials in the microwave frequency range were achieved by using ferroelectrics, ferri-magnetics and ferro-magnetics [26–28]. LCs are a good platform for achieving tunability in nearly all frequency ranges, and for microwaves the tunable metamaterials were demonstrated by using either the electric or magnetic fields [29, 30]. This method is, however, often impractical for bulk microwave metamaterial use due to large amounts of expensive LC required. A completely new class of tunable (and nonlinear) metamaterials was developed after detailed studies of the near-field interaction between metamaterial elements have shown strong potential and importance of this method [15]. Mechanical displacements of the metamaterial elements lead to change in the coupling of these elements, and this tunes electromagnetic response of metamaterials. This was demonstrated both in microwave and THz frequency ranges, and in principle possibly in optics, though required tolerances make it technologically difficult. An extension of this concept involves the use of MEMS and pneumatic metamaterials, where the displacement of metamaterial elements is achieved either by activating the MEMS mechanisms or by changing pressure [11, 31].

6.3.1 Tunable terahertz metamaterials

Metamaterials are well placed for applications in manipulating THz waves, since there are a small number of natural materials that can be used for these purposes. The first demonstration of tunable THz metamaterials was done in 2006, when the Schottky-type semiconductor structure with metamaterial-shaped electrodes exhibited strong tunability by voltage [32]. In such a structure, the conductivity of the substrate was changing, resulting in the modification of the strength of the metamaterial resonance. Further advances in this field were done by creating a patterned semiconductor structure that would have conductivity increased only in certain areas of metamaterial, leading to the frequency shift of the metamaterial resonance [33, 34]. In these structures, the change in the semiconductor excitation was achieved via photo-doping by intense laser beam.

Ferroelectrics [35] as well as vanadium dioxide [36] can be used for creating temperature-tunable metamaterials. At the same time, it was shown that the temperature for tuning LC in filtrated THz metamaterials is not efficient [37]. The general concept of tuning metamaterial response by displacing meta-atoms can be readily applied in the THz frequency range. This was achieved by direct mechanical motion of the metamaterial layers [38] by using polymer actuation [39] by pneumatic actuation [40] or by stretching [41]. We will focus on the first two of these methods along with the description of the liquid metamaterial, whose properties change when we apply a dc electric field [42] that rotates the metamolecules.

6.3.2 Mechanically shifting of metamaterial layers

In order to illustrate the significant alteration of the propagation spectrum in a double-layer metamaterial, electrically coupled metamaterial layers that have resonators has been examined [38]. Because of the electrical connections between the resonators in adjacent cells, the structure only shows significant transmission in the vicinity of its resonances. Conventional optical lithography is used to create the metamaterial layers, and then a lift-off procedure is performed. The resonators are arranged in a square lattice with a lattice constant of 135 μm, and the gold has a thickness of 150 nm. The overall area of the structure is 5 mm × 5 mm. Following the fabrication of two metamaterial samples, each sample is covered with a small quantity of heat-sensitive glue that has solidified and is heated until it reaches the melting point. Under a microscope, the two metamaterial layers are face-to-face bonded and aligned with one another. By warming up the sample and shifting the upper layer relative to the lower layer, one can adjust the lateral displacement. By shifting the upper layer back and forth to spread the adhesive, the distance between the two layers can be adjusted. Tunability is obtained by adjusting the coupling between the two neighboring metamaterial structures; the glue's thickness controls the maximum frequency shift, while the structures' lateral movement controls the frequency shift. The lateral displacements from half unit cell $S = 67.5$ μm to $S = 0$ μm can be changed. Numerical simulations are performed using a commercial electromagnetic solver called CST Microwave Studio. In the simulations, the authors have used normal incident waves polarized across the gap of the resonator. It is expected that the boundary layers in the z-direction exactly match to terminate the computing region, and that the structures are periodic in the x–y-plane. The x–y-plane structures are thought to be periodic, and the boundary layers in the z-direction exactly match the end of the computing region. Six places around the metamaterial region were measured, and the average spacing was 10.4 μm, with variances below 2.5 μm. An apparatus for measuring THz time-domain spectroscopy is used to measure transmission through the double-layered metamaterials. The incident field's polarization is parallel to the resonator's gaps. The system is excited by a femtosecond laser pulse, and a time domain sample of the transmission response is acquired. The transmitted wave's amplitude and phase are obtained via a Fourier transform, and the experimental transmission spectrum can be computed through additional frequency-domain signal processing. Time gating in the time-domain signal reduces Fabry–Perot interference fringes in the transmission spectrum.

Figure 6.2. displays the results of the simulation and experimentation for the double-layered structure with various lateral displacements.

Additionally, alignment for various lateral displacements is displayed. The suggested post-processing approach is feasible, as seen by the reasonably excellent alignment obtained by manually adjusting alignment markings under a microscope. There is a significant shift in the transmission band, and the spectrum is reshaped as a result. The scientists also computed the resonance shift for a greater 110 μm gap between the layers. The vertical connection between two layers is greatly diminished at such a big distance. In this instance, the resonance of stacked layers is like that of isolated metamaterial layers because there aren't any strong interactions between them. Because of this, the lateral movement in this instance will result in very little resonance tuning. This post-processing technique can result in a large tunability in two layers and may find use in the construction of tunable THz devices without

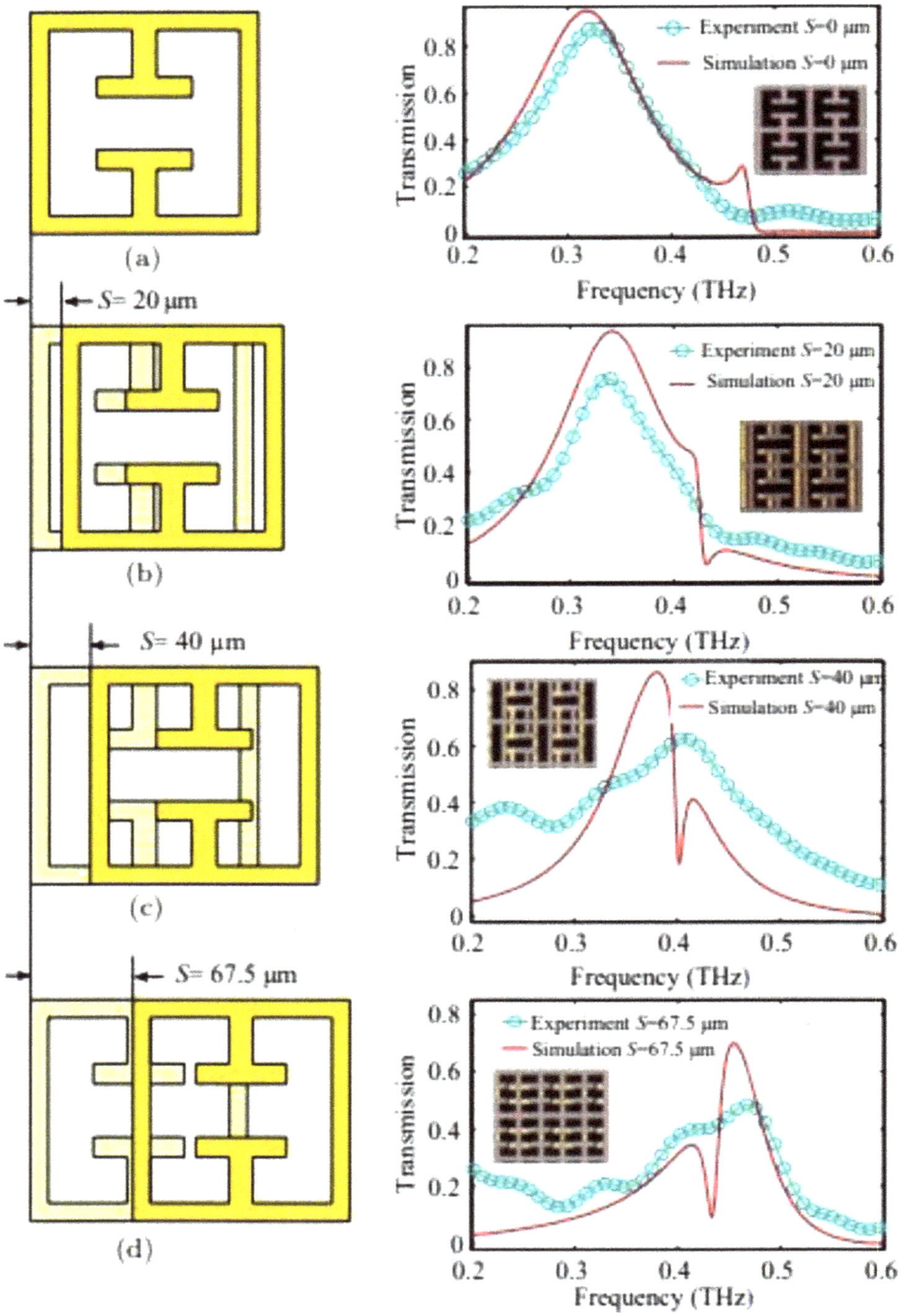

Figure 6.2. Diagrams, comparison of simulation results with measurements, and alignment of the double-layered structure under a microscope (inset) (a) $S = 0$ µm, (b) $S = 20$ µm, (c) $S = 40$ µm, (d) $S = 67.5$ µm. Adapted from [38], with permission from AIP Publishing.

requiring re-fabrication procedures. A THz filter composed of two metamaterial layers—one permanent and the other movable via a translation stage—can be imagined. Other kinds of metamaterial patterns can be used to design different spectral features.

6.3.3 Conductive polymers for metamaterial tuning

Conductive polymers can be used to accomplish electric control of the metamaterial layers offset [39]. Conducting polymers, also known as π-conjugated polymers, have been extensively researched for their potential uses in the development of organic light-emitting diodes, organic thin-film transistors (TFTs), organic photovoltaic cells, and other electronic devices. They have also been used to develop soft actuators. The basic working principle typical in π-conjugated polymer actuators is an electrochemical reduction/oxidation reaction, in which doped ions move into or out of the π-conjugated polymer by applying and reversing voltages, giving mechanical deformation of the polymer in a reversible manner. It was found that heavily doped polypyrrole (PPy) film could be contracted by applying only a few Volts, which controls the desorption of water molecules. Such PPy films are known to behave as metals in the THz and lower frequency ranges and can exhibit extraordinary transmission in a perforated film [36].

The authors have showcased an electrically adjustable apparatus operating within the THz frequency range. Nevertheless, their methodology might potentially be expanded to encompass near-infrared or visible wavelengths. Through electrochemical polymerization, they have created thin films of PPy doped with hexafluorophosphate (PF6) ions, or PPy (PF6). In order to laterally change the relative positions of two stacked metamaterial arrays intended to function as electro-active THz meta-devices, they are employed as linear actuators. The metamaterials design as seen in the preceding section was employed in this experiment, with one resonator array stationary and the next array allowed to move laterally. Due to the elastic nature of the film, the PPy (PF6) linear actuators are coupled to both sides of the freely moveable metamaterial array. This allows for one-dimensional (1D) linear lateral shift along the capacitive gap of the SRR as well as the recovery of the starting position. Since PPy (PF6) films are said to contract only a small percentage, we decided to cut the film into square pieces measuring 10 mm by 10 mm. This will allow us to reach an actuation range of tens of μm, which will enable us to gain a considerable frequency shift in our THz device. Since the electric field-coupled resonators that make up each metamaterial layer's repeating unit are all connected, the single-layer structure only exhibits resonant transmission near its resonances, which occurs at 0.375 THz. It has been confirmed that by stacking these CESRR arrays, two resonant modes were observed at 0.318 and 0.468 THz. The in-phase and out-of-phase loop currents are represented by these two modes, which are symmetric and anti-symmetric, respectively [37]. Because of the symmetric mode, the well-aligned device exhibits resonance at around 0.320 THz (figure 6.3(a)). By providing a small bias voltage of just a few volts to one of the polymer actuators, the researchers were able to pull the SRR array in a single direction and achieve a

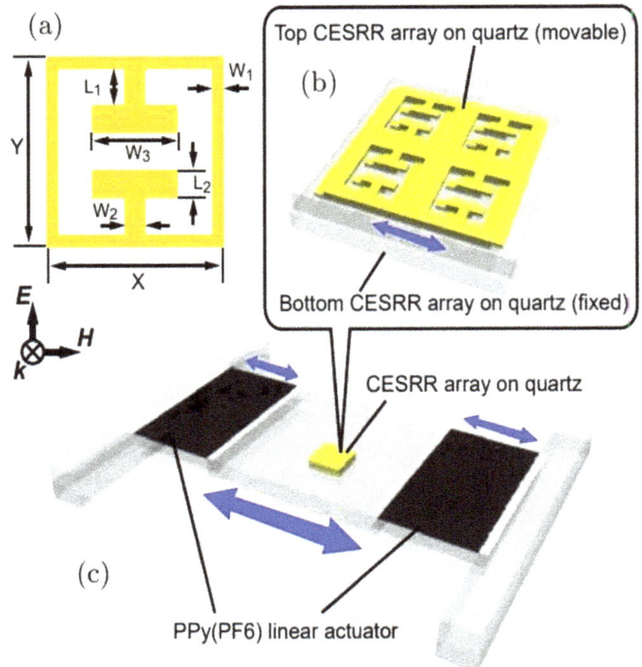

Figure 6.3. Schematics of the tunable THz device based on a PPy (PF6) linear actuator. (a) Geometry of the unit cell along with the polarization directions of the incident THz wave (top view). (b) Schematics of the double-layered metamaterial arrays with fixed substrate and movable superstrate. (c) Schematics of the tunable THz device based on double-layered metamaterial with fixed substrate and movable superstrate suspended with PPy (PF6) linear actuator. The top layer can be laterally moved by applying voltage. Adapted from [39], with permission from AIP Publishing.

frequency shift of resonant THz transmission. The polymer film contracts when voltage is applied, shifting the freely moveable layer laterally. The transmission spectrum and starting condition of the system are restored by cutting off the applied voltage. The optical microscope images also confirm that an applied voltage of 1.5 V is enough to have significant lateral shift of the SRR array to obtain frequency shift in resonant transmission in our device, as shown in figure 6.3(b). For MEMS-type devices, this method can be a low-cost substitute for a more involved and costly fabrication procedure. The measured and numerically simulated results have been shown in figure 6.4.

6.3.4 Meta liquid crystals (MCL)

Meta liquid crystals (MLCs) are another structure that takes advantage of the idea of the metamolecules' mechanical motion to adjust its properties [42]. The meta-atoms are enclosed in elongated dielectric bars that are scattered throughout a host liquid and rotate due to electrophoretic forces in response to an applied biasing electric field to achieve tunability. When all of these particles is oriented parallel to

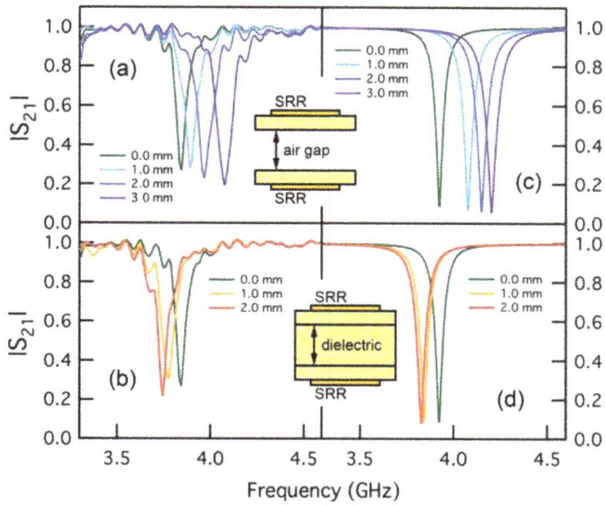

Figure 6.4. (a, b) Measured and (c, d) numerically simulated transmission coefficients of the stacked SRR on dielectric substrate with varying spacing between them with air gap (a, c) and with dielectric in between (b) and (d). Adapted from [39], with permission from AIP Publishing.

the field, the structure will show strong anisotropic characteristics. The direction of the biasing field can be changed to rotate the anisotropy axis, allowing the metamaterial to be tuned electromagnetically in a way akin to liquid crystal mesogens. The described metamaterial is referred to as MLC, or constituent elements meta-mesogens, because of this powerful resemblance. Complex meta-atoms with sub-wavelength characteristics make up MLCs, and a wide range of tailored properties are possible. The researchers would also create MLCs with tunable properties not possible with conventional LCs, like larger anisotropy, strong magnetic and/or chiral (and generally an isotropic) responses, or well-designed spectral features like electromagnetically induced transparency, because the electrostatic and electromagnetic properties of our meta-mesogens are weakly correlated. Such metamaterials' liquid state makes it possible for them to flow over surfaces of any 3D shape and fill in gaps, thus expanding the range of applications for which they can be used. Although the approach might be implemented across a broad frequency range, we here experimentally illustrate it within the THz frequency range. The conventional liquid crystal and researchers developed meta liquid crystals are shown in figure 6.5.

Because natural materials lack intrinsic responsiveness, metamaterials are essential for regulating THz waves and constructing passive, tunable electronics. The researchers selected two varieties of meta-atoms: I-beam resonators (IBRs) and electric split-ring resonators (ESRRs). The entire meta-mesogen behaves as an anisotropic dipole in an external bias field because each meta-mesogen is made up of a brief array of identical metallic units encased in a 10-μm thick polyimide. Since the selected meta-atoms exhibit a strong anisotropic response to the high frequency field, the composite medium's overall response can be tuned by rotating the entire meta-mesogen.

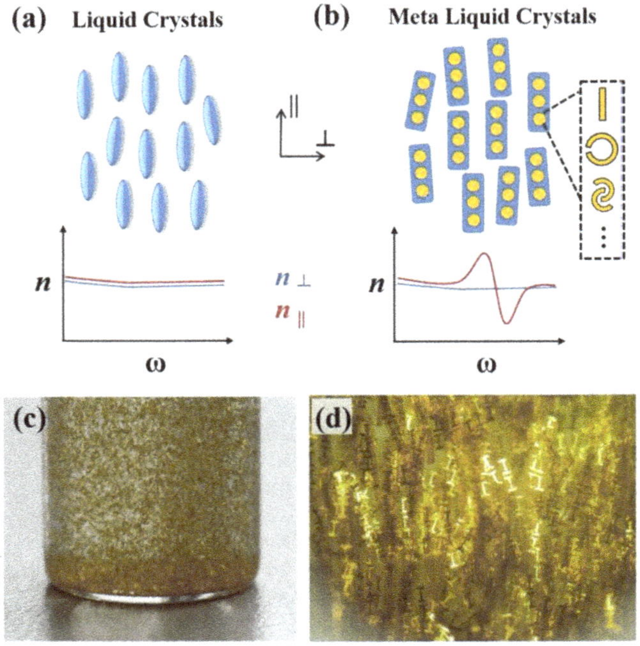

Figure 6.5. Schematics and corresponding anisotropic refractive index components of (a) a conventional liquid crystal in the nematic phase and (b) meta-liquid crystals with meta-mesogens oriented along the external static electric field. The electromagnetic properties of the MLC are defined by meta-atoms, while the size and shape of the meta-mesogens largely define the response to the control electric field. (c) Photograph of the fabricated liquid metamaterial in a glass bottle. (d) Microscopic image of meta-mesogens aligned under external bias electric field. Adapted from [42], with permission from AIP Publishing.

Three designs have been examined by the researchers: two IBRs (2-IBR), five ESRRs (5-ESRR), and three ESRRs (3-ESRR) combined to form a meta-mesogen. In order to allow each type of meta-mesogen to freely rotate in response to an external bias electric field, they are combined in a non-polar liquid. Since paraffin oil is a widely accessible, safe-to-handle liquid with a reasonable degree of transparency in the THz region, it is used in the experimental work.

The dipole approximation provides a good description of the electrostatic torque of a meta-mesogen in a uniform static electric field. The length and aspect ratio of the meta-mesogens play a major role in this process. The torque can be used to determine the meta-mesogens' response time in various liquids. For a bias field strength of 1.6×10^5 V m^{-1}, all three designs exhibit an estimated response time of approximately 100 ms in paraffin oil. It is possible to improve this time even further by utilizing less viscous liquids. Furthermore, because paraffin oil sinks more slowly than mesogens and is less dense than them, it is not the best material for achieving MLC. It is necessary to look for more appropriate host liquids; as an example, the authors used trichloroethylene to simulate MLC. It is better suited for a faster operating MLC because it is denser and less viscous than paraffin oil. This might increase the operation speed by at least one order of magnitude, as the computations indicated. Trichlorethylene was not used in the experiments due to its high toxicity. The dynamics can further be optimized by choosing the shape and size of the meta-mesogens. Since the sign of the voltage output from our source cannot be changed, there is a net dc contribution leading to charge build up, which would result in the aggregation of meta-mesogens, and therefore bursts of pulses are used to maintain the alignment. To enable practical application of this idea, further technical optimization is required so that the charging effect and sinking problem can be addressed.

6.4 Some reconfigurable FSS using metamaterial

Here, various reconfigurable frequency selective surface structures based on metamaterials, researched in recent years, have been described.

Dual-band dynamic FSS using origami (2020): The art of origami, which involves folding sheets of paper into three-dimensional shapes, has recently attracted the attention of scientists and engineers from a variety of fields. Over the past ten years, novel designs of mechanical metamaterials have been introduced, utilizing their geometrical and structural features. Electromagnetic waves have been shaped and controlled by the application of frequency selective surfaces (FSSs). The intricate geometry used in the antiquated design techniques are difficult to apply. The researchers altered the forms of FSS designs based on origami patterns in order to produce improved electromagnetic performance and new degrees of freedom. This was done with the intention of transforming electromagnetic waves. Strong coupling electromagnetic resonators with origami patterns has been developed in this experiment [43]. Researchers have transformed a single-band FSS to a dual-band FSS. They also explained this transformation by showing that both symmetric and anti-symmetric modes are excited due to the strong coupling and suitable orientation of the elements. The suggested origami FSS could be folded and unfolded, which allows it to adjust or reconfigure its dual-band performance. As a result, unlike typical FSSs, which are static and unable to alter their performance, the suggested FSS is a dynamic changeable electromagnetic structure.

The folded structure of the origami-based FSS for different folding angles and their corresponding frequency characteristics is shown in figure 6.6.

A tunable metamaterial frequency selective surface with variable modes of operation (2009): In 2009, a miniaturized frequency selective surface (FSS) with changeable elements was introduced. The design's performance was assessed using a typical waveguide measurement setup. A very thin **dielectric** substrate was sandwiched between two periodic arrays of metallic loops with the same periodicity that made up the proposed FSS. Numerical evidence has demonstrated that the reconfigurable surface can be tuned by integrating tuning varactors into the structure. There is now a reconfigurable frequency response with two modes of operation: bandstop and bandpass, thanks to varactors on both layers. The response's bandwidth and center frequency were adjusted separately, in addition to its two entirely distinct modes of operation. A continuous bandwidth over a range of 3–3.5 GHz was provided by the suggested FSS [44]. This study presented a

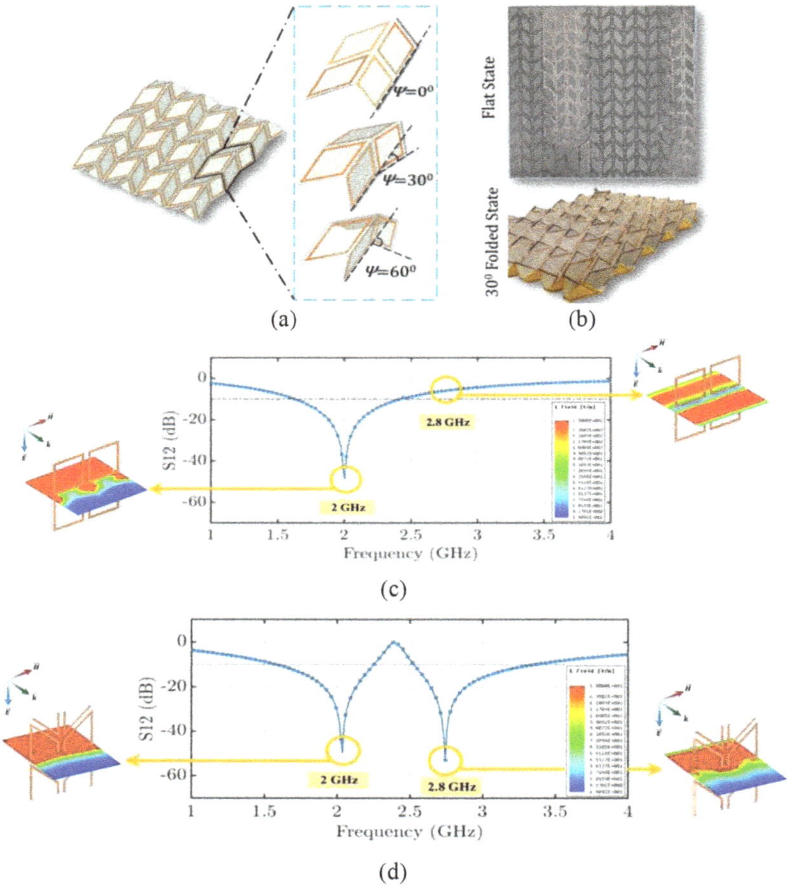

Figure 6.6. Rhombic loop FSS on a Miura-Ori geometry. (a) 4 × 3 FSS film in an arbitrary folded angle showing the unit cell at different folding states. From top to bottom: flat unit cell at folded angle of 0°, folded unit cell at folded angle of 30°. and folded unit cell at folded angle of 60°. (b) Fabricated FSS on polyimide substrate in its flat state (top) and folding state for folded angle of 30° (bottom). Electric field distribution for TE mode excitation for: (c) folded angle of 0° corresponding to standard square loop FSS, which exhibits single-band filtering performance and (d) Corresponding folded angle of 60° to the proposed origami FSS, which exhibits dual-band filtering performance. Reproduced from [43]. CC BY 4.0.

reconfigurable miniaturized-element FSS, in which the response may be converted from bandpass to stopband, in addition to the bandwidth and center frequency being able to be changed independently. This FSS is printed on both sides of a thin, flexible substrate with two arrays of loops. By using varactors to connect the loops on each layer, the architecture can be made more tunable. The design process and a physical understanding of the behavior of the structure are first described below. It will be demonstrated that this FSS's response consists of two transmission zeros, or two degrees of freedom, allowing for extremely flexible frequency adjustment. The validity of the design process is subsequently demonstrated by comparing the waveguide measurement results with the numerical calculations (figure 6.7).

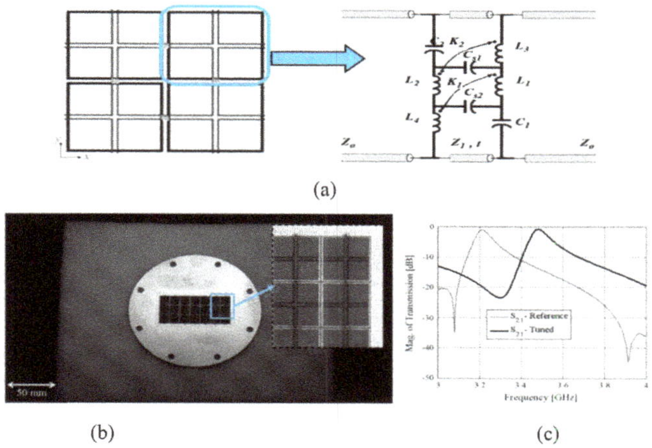

Figure 6.7. (a) Reconfigurable FSS. Four-unit cells are shown; the gray loop array and the black one is printed on opposite sides of the substrate. Varactors are marked by zebra-striped squares (crossing points) and circuit model of the reconfigurable FSS consisting of two parallel notch circuits that are coupled with each other through magnetic effects and capacitive junctions. (b) Waveguide prototype of the reconfigurable FSS compatible with WR-284 standard flan. (c) Free-space simulations of the reconfigurable FSS using periodic boundary condition setup in HFSS at normal incidence compared to ADS's circuit simulations. Reproduced from [48]. CC BY 4.0.

Reconfigurable all-dielectric metamaterial frequency selective surface based on high-permittivity ceramics (2016): The creation of reconfigurable all-dielectric metamaterial frequency selective surfaces (FSSs) devoid of metallic components was suggested by the researcher in this study, utilizing high-permittivity ceramics. The size of the ceramic particles is sub-wavelength because of their high permittivity. Thus, effective permittivity and permeability can be used to describe the proposed FSS's macroscopic electromagnetic properties in an equivalent way. The FSS's EM response can alternate between two nearby bands by altering the ceramic resonators' orientation. A reconfigurable FSS made up of cross-shaped resonators was created, manufactured, and measured in order to validate the design process. By modifying the geometrical parameters, the resonant frequencies of the ceramic particles can be set to produce two contiguous stop bands and to ensure that the first two resonant modes are correctly spaced. The FSS can switch between a 0.4 GHz stopband in 4.54–4.94 GHz and a 1.0 GHz stopband in 3.55–4.60 GHz. The bandstop property of the developed FSS is verified by both simulation and experiment findings. Such FSSs avoid the low breakdown voltages and ohmic losses associated with metallic structures because they use high-permittivity ceramic particles as the unit cell. As a result, these FSSs have several benefits in high-power and high-temperature applications. By rotating a single structure, this type of reconfigurable FSS can create two neighboring stop bands. It can be applied to reconfigurable antenna systems and RCS reduction approaches, among other things.

Reconfigurable frequency selective surface based on atmospheric light intensity (2019): In [49] Saha *et al* reported a metamaterial-based reconfigurable FSS with tunable characteristics depending on the atmospheric light intensity. The unit cell

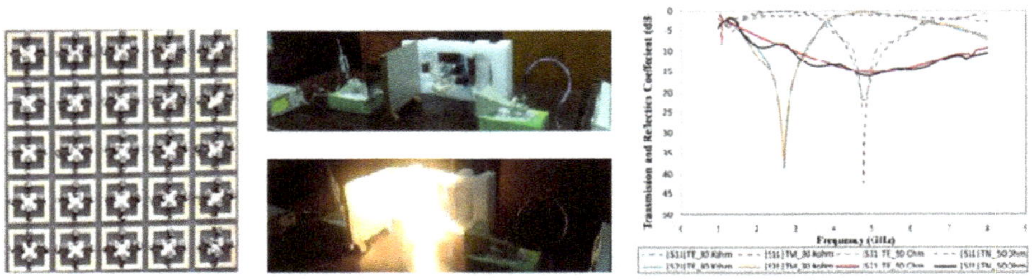

(a) (b) (c)

Figure 6.8. (a) LDR-based reconfigurable FSS structure. (b) Measurement set up on microwave test bench for dark and bright conditions of atmospheric light. (c) Transmission characteristics (S21) of the proposed design for both conditions. Reproduced from [48].

structure was formed by a metallic square patch surrounded by metallic square loop. The light dependent resistors (LDRs) have been used as atmospheric light sensors. Variations in the light intensity that falls upon it are used as the control parameter for variations in the resistance values of LDRs, resulting in tunable transmission characteristics of the FSS. The simulated results demonstrate that the FSS has bandpass properties at the frequency range of C-band (4–8 GHz) and bandstop properties at the frequency range of lower S-band (2–3 GHz) in the absence of atmospheric light and stopband properties at the frequency range of C-band (4–8 GHz) in the presence of full bright light. In figure 6.8, the fabricated structure, measurement setup and reconfigurable transmission behavior of the proposed structure are shown clearly.

6.5 Reconfigurable antenna using metamaterial based FSS

Metamaterials are materials that are often created to provide electromagnetic properties that are uncommon or challenging to achieve in nature. These materials typically have new or artificial structures. Metamaterials have generated significant interest and may find utility in various electromagnetic applications, ranging from microwave to optical regimes, due to their potential to offer engineerable permittivity, permeability, and index of refraction (η). This section of the chapter provides a thorough overview of the most current investigations into those metamaterial-based reconfigurable antennas. A wide variety of antenna examples are included to facilitate the understanding for general readers.

As the demand for flexible designs grows, various techniques have been employed to address the need for reconfigurable and flexible metamaterial designs. Metasurfaces are engineered surfaces that exhibit unique properties not found in nature. This chapter explores flexible metamaterial designs implemented using microfluidics to achieve reconfigurability. Microfluidics is the technology of manipulating fluids of small volumes, typically within 1 nL to 1 aL, within channels of tens to hundreds of micrometers to control chemical, biological and physical processes [45, 46]. Recently, microfluidic techniques have been employed to develop devices such as lab on chip designs, point-of-care diagnostic devices, and drug development kits, as a result of their compactness, low energy consumption, and ease of fabrication and use. The simplest form of a microfluidic device comprises fluids typically enclosed in optically transparent substrates such as glass, silicon, polymer, PDMS, and actuators. The actuation method used to pump or drive the fluids in the micro-channels includes piezoelectric, electrostatic, electromagnetic, electrophoretic, electro-osmotic, bimetallic and shape memory alloy [47]. Microfluidic devices can be categorized into two forms based on the manipulation and handling of the fluid flow: continuous-flow microfluidics and digital (droplet-based) microfluidics [48]. This chapter explores both techniques towards developing reconfigurable antenna designs.

Metamaterial-based frequency reconfigurable microstrip antenna for wideband and improved gain performance: A metamaterial-based frequency reconfigurable micro-strip antenna is proposed with wideband tuning range and enhanced gain. By co-designing the main reconfigurable microstrip patch unit and the upper-layer located reconfigurable negative-permeability-metamaterial resonant loop, the gain of the resultant proposed antenna is readily improved. The wideband reconfigurability of the microstrip antenna and the metamaterial unit is realized by tuning the capacitance values of the varactor diodes through altering their bias voltages. Furthermore, the effect of different numbers of reconfigurable metamaterial units on the gain of the proposed antenna is extensively explored, which shows the radiation pattern is largely improved since the electromagnetic energy is more concentrated and directed. Finally, the proposed antenna is fabricated and measured. The experimental results demonstrate that the operating range of the proposed antenna is within 8.6–10.3 GHz, and the reflection coefficient is less than −10 dB, and the gain is above 6 dBi. The radiation efficiency of the antenna is above 85%. The electrical size of the antenna is $0.58\lambda0 \times 0.58\lambda0$. As expected, the proposed antenna has the main advantage of enhanced gain performance, while performing compa-rable frequency tuning range [50]. In figure 6.9, the proposed structure of the reconfigurable antenna and its tunable reflection coefficient (S11) for different values of capacitance in varactor diodes are shown clearly.

Metamaterial-based, vertically polarized, miniaturized beam-steering antenna for reconfigurable sub-6 GHz applications: A miniaturized, vertically polarized (VP), pattern switchable/beam-steering antenna based on a metamaterial radiator is presented for sub-6GHz applications [51]. Different from the traditional split-ring resonator (SRR), the proposed T-shaped SRR can work as a miniaturized VP omnidirectional radiator. To steer the radiation beam, an oval switchable mush-room reflector with a low profile and low loss is explored and utilized. Multiple switchable mushroom reflectors are loaded around the T-shaped SRR radiator to obtain various steerable beams. To verify the proposed design principle, a VP beam-steering SRR-based antenna with a compact size of $0.36\lambda_0 \times 0.36\lambda_0 \times 0.07\lambda_0$ is designed by loading four mushrooms. By controlling the off/on states of four switches, the proposed reconfigurable antenna can realize beam-steering in the horizontal plane (eight states with 45°-step). The measured overlapped −10 dB impedance bandwidth is wider than 3.40–3.80 GHz (11.1%), and the realized peak gain is larger than 5.3 dBi. Therefore, the proposed antenna is a good solution for

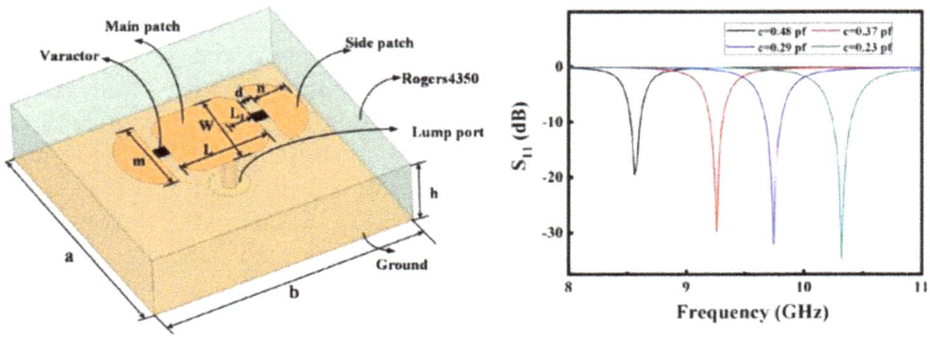

(a) (b)

Figure 6.9. (a) Metamaterial structure-based antenna design prototype using varactor diode as active element. (b) Simulated tunable frequency characteristics of S11 of the proposed antenna. [50] John Wiley & Sons. [© 2021 Wiley Periodicals LLC.]

the widely deployed fifth generation (5G)-NR and smart antenna applications due to its compact size, flexible beam-steering capacity, broad bandwidth, and good scanning coverage [51].

In this chapter, a comprehensive review on the experimental research on metamaterial-based reconfigurable antenna and frequency selective surface usable in modern communication systems is discussed. The application of metamaterials includes the enhancement of bandwidth and gain, size reduction, efficiency enhancement, reconfiguration of frequency and pattern, etc. For bandwidth, gain, and efficiency improvement, the metamaterials are loaded in the front or back of the reference antenna to obtain the desired improved results. In the aforementioned discussion, examples from the literature are given to show the workings and results of metamaterial-loaded antennas. For size reduction, the metamaterial is placed on the same plane as the antenna to obtain a miniaturized antenna with the same results as a larger antenna. Afterward, the reconfiguration application of metamaterials is discussed in terms of frequency reconfigurability and pattern reconfigurability, along with various examples from the literature. Then, the current trend of metamaterials is discussed in terms of its applications in intelligent selective surfaces and absorbers. Researchers propose a number of studies, but only a few interesting articles are reported in this chapter to serve as examples and evidence.

References

[1] Zheludev N I and Kivshar Y S 2012 From metamaterials to metadevices *Nat. Mater.* **11** 917–24

[2] Lapine M, Powell D, Gorkunov M, Shadrivov I, Marqués R and Kivshar Y 2009 Structural tunability in metamaterials *Appl. Phys. Lett.* **95** 084105

[3] Karim M F, Liu A Q, Alphones A and Yu A B 2006 A tunable bandstop filter via the capacitance change of micromachined switches *J. Micromech. Microeng.* **16** 851

[4] Gil I, Martın F, Rottenberg X and De Raedt W 2007 Tunable stop-band filter at Q-band based on RF-MEMS metamaterials *Electron. Lett.* **43** 1153–4

[5] Hand T and Cummer S 2007 Characterization of tunable metamaterial elements using MEMS switches *IEEE Antennas Wirel. Propag. Lett.* **6** 401–4

[6] Ozbey B and Aktas O 2011 Continuously tunable terahertz metamaterial employing magnetically actuated cantilevers *Opt. Express* **19** 5741–52

[7] Pryce I M, Aydin K, Kelaita Y A, Briggs R M and Atwater H A 2010 Highly strained compliant optical metamaterials with large frequency tunability *Nano Lett.* **10** 4222–7

[8] Aksu S, Huang M, Artar A, Yanik A A, Selvarasah S, Dokmeci M R and Altug H 2011 Flexible plasmonics on unconventional and nonplanar substrates *Adv. Mater.* **23** 4422–30

[9] Kasirga T S, Ertas Y N and Bayindir M 2009 Microfluidics for reconfigurable electromagnetic metamaterials *Appl. Phys. Lett.* **95** 214102

[10] Ou J Y, Plum E and Zheludev N I 2012 MHz bandwidth electro-optical modulator based on a reconfigurable photonic metamaterial *CLEO: Applications and Technology* pp JW4A-11 (Washington, DC: Optica Publishing Group)

[11] Liu A Q, Zhu W M, Tsai D P and Zheludev N I 2012 Micromachined tunable metamaterials: a review *J. Opt.* **14** 114009

[12] Zhong J, Huang Y, Wen G, Sun H and Zhu W 2012 The design and applications of tunable metamaterials *Procedia Eng.* **29** 802–7

[13] Turpin J P, Bossard J A, Morgan K L, Werner D H and Werner P L 2014 Reconfigurable and tunable metamaterials: a review of the theory and applications *Int. J. Antennas Propag* **2014** 29837

[14] Zheludev N I and Plum E 2016 Reconfigurable nanomechanical photonic metamaterials *Nat. Nanotechnol.* **11** 16–22

[15] Powell D A, Lapine M, Gorkunov M V, Shadrivov I V and Kivshar Y S 2010 Metamaterial tuning by manipulation of near-field interaction *Phys. Rev.* B **82** 155128

[16] Li L, Wang J, Wang J, Ma H, Du H, Zhang J and Xu Z 2016 Reconfigurable all-dielectric metamaterial frequency selective surface based on high-permittivity ceramics *Sci. Rep.* **6** 24178

[17] Li A, Singh S and Sievenpiper D 2018 Metasurfaces and their applications *Nanophotonics* **7** 989–1011

[18] Salisbury W W 1952 U.S. Patent No. 2,599,944 (Washington, DC: U.S. Patent and Trademark Office)

[19] Lim S, Caloz C and Itoh T 2004 Metamaterial-based electronically controlled transmission-line structure as a novel leaky-wave antenna with tunable radiation angle and beamwidth *IEEE Trans. Microw. Theory Tech.* **52** 2678–90

[20] Gil I, Bonache J, Garcia-Garcia J and Martin F 2006 Tunable metamaterial transmission lines based on varactor-loaded split-ring resonators *IEEE Trans. Microw. Theory Tech.* **54** 2665–74

[21] Shadrivov I V, Morrison S K and Kivshar Y S 2006 Tunable split-ring resonators for nonlinear negative-index metamaterials *Opt. Express* **14** 9344–9

[22] Shadrivov I V, Kapitanova P V, Maslovski S I and Kivshar Y S 2012 Metamaterials controlled with light *Phys. Rev. Lett.* **109** 083902

[23] Kapitanova P V, Slobozhnanyuk A P, Shadrivov I V, Belov P A and Kivshar Y S 2012 Competing nonlinearities with metamaterials *Appl. Phys. Lett.* **101** 231904

[24] Ricci M C, Xu H, Prozorov R, Zhuravel A P, Ustinov A V and Anlage S M 2007 Tunability of superconducting metamaterials *IEEE Trans. Appl. Supercond.* **17** 918–21

[25] Castellanos-Beltran M A, Irwin K D, Hilton G C, Vale L R and Lehnert K W 2008 Amplification and squeezing of quantum noise with a tunable Josephson metamaterial *Nat. Phys.* **4** 929–31

[26] Hand T H and Cummer S A 2008 Frequency tunable electromagnetic metamaterial using ferroelectric loaded split rings *J. Appl. Phys.* **103** 066105
[27] He Y, He P, Yoon S D, Parimi P V, Rachford F J, Harris V G and Vittoria C 2007 Tunable negative index metamaterial using yttrium iron garnet *J. Magn. Magn. Mater.* **313** 187–1
[28] Kang L, Zhao Q, Zhao H and Zhou J 2008 . Magnetically tunable negative permeability metamaterial composed by split ring resonators and ferrite rods *Opt. Express* **16** 8825–34
[29] Zhang F, Zhang W, Zhao Q, Sun J, Qiu K, Zhou J and Lippens D 2011 Electrically controllable fishnet metamaterial based on nematic liquid crystal *Opt. Express* **19** 1563–8
[30] Zhang F, Kang L, Zhao Q, Zhou J, Zhao X and Lippens D 2009 Magnetically tunable left handed metamaterials by liquid crystal orientation *Opt. Express* **17** 4360–6
[31] Khodasevych I E, Shadrivov I V, Powell D A, Rowe W S T and Mitchell A 2013 Pneumatically switchable graded index metamaterial lens *Appl. Phys. Lett.* **102** 031904
[32] Chen H T, Padilla W J, Zide J M, Gossard A C, Taylor A J and Averitt R D 2006 Active terahertz metamaterial devices *Nature* **444** 597–600
[33] Chen H T, O'hara J F, Azad A K, Taylor A J, Averitt R D, Shrekenhamer D B and Padilla W J 2008 Experimental demonstration of frequency-agile terahertz metamaterials *Nat. Photonics* **2** 295–8
[34] Kafesaki M, Shen N H, Tzortzakis S and Soukoulis C M 2012 Optically switchable and tunable terahertz metamaterials through photoconductivity *J. Opt.* **14** 114008.
[35] Němec H, Kužel P, Kadlec F, Kadlec C, Yahiaoui R and Mounaix P 2009 Tunable terahertz metamaterials with negative permeability *Phys. Rev.* B **79** 241108
[36] Driscoll T, Kim H T, Chae B G, Kim B J, Lee Y W, Jokerst N M and Basov D N 2009 Memory metamaterials *Science* **325** 1518–21
[37] Liu L, Shadrivov I V, Powell D A *et al* 2013 Temperature control of terahertz metamaterials with liquid crystals *IEEE Trans. Terahertz Sci. Technol.* **3** 827–31
[38] Liu L, Chen W C, Powell D A *et al* 2014 Post-processing approach for tuning multi-layered metamaterials *Appl. Phys. Lett.* **105** 151102
[39] Matsui T, Inose Y, Powell D A and Shadrivov I V 2016 Electroactive tuning of double-layered metamaterials based on π-conjugated polymer actuators *Adv. Opt. Mater.* **4** 135–40
[40] Kan T, Isozaki A, Kanda N, Nemoto N, Konishi K, Takahashi H and Shimoyama I 2015 Enantiomeric switching of chiral metamaterial for terahertz polarization modulation employing vertically deformable MEMS spirals *Nat. Commun.* **6** 8422
[41] Li J, Shah C M, Withayachumnankul W, Ung B S Y, Mitchell A, Sriram S and Abbott D 2013 Mechanically tunable terahertz metamaterials *Appl. Phys. Lett.* **102** 121101
[42] Liu M, Fan K, Padilla W, Powell D A, Zhang X and Shadrivov I V 2016 Tunable meta-liquid crystals *Adv. Mater.* **28** 1553–8
[43] Biswas A, Zekios C L and Georgakopoulos S V 2020 Transforming single-band static FSS to dual-band dynamic FSS using origami *Sci. Rep.* **10** 13884
[44] Bayatpur F and Sarabandi K 2009 A tunable metamaterial frequency-selective surface with variable modes of operation *IEEE Trans. Microw. Theory Tech.* **57** 1433–8
[45] Stroock A D 2008 Microfluidics *Optical Biosensors* 2nd edn ed F S Ligler and C R Taitt (Amsterdam: Elsevier) 659–81
[46] Whitesides G M 2006 The origins and the future of microfluidics *Nature* **442** 368–73
[47] Cho S K, Moon H and Kim C J 2003 Creating, transporting, cutting, and merging liquid droplets by electrowetting-based actuation for digital microfluidic circuits *J. Microelectromech. Syst.* **12** 70–80

[48] Nguyen N T, Hejazian M, Ooi C H and Kashaninejad N 2017 Recent advances and future perspectives on microfluidic liquid handling *Micromachines* **8** 186
[49] Saha S, Begam N, Chatterjee A, Biswas S and Sarkar P P 2019 Reconfigurable frequency selective surface with tunable characteristics depending on intensity of atmospheric light *IET Microw. Antennas Propag.* **13** 2336–41
[50] Zhou D, Wang H, Deng L, Qiu L-L and Huang 2022 Metamaterial-based frequency reconfigurable microstrip antenna for wideband and improved gain performance *Int. J. RF Microw. Comput. Aided Eng.* **32** 1–8
[51] Wang Z and Dong Y 2022 Metamaterial-based, vertically polarized, miniaturized beam-steering antenna for reconfigurable sub-6 GHz applications *IEEE Antennas Wirel. Propag. Lett.* **21** 2239–43

Chapter 7

Design of metamaterial-based antennas and FSSs for biomedical applications

This chapter deals with various applications of metamaterial-based antennas as well as frequency selective surfaces (FSSs) in the field of biomedical engineering. The sub-disciplines of biomedical engineering comprise development of implantable medical devices, biomedical telemetry, medical imaging, tissue engineering, and biomedical signal processing [1–4]. Among these, biomedical telemetry primarily focuses on the application of telemetry in healthcare to remotely monitor different medical symptoms and health parameters of patients [5, 6]. Antennas as well as frequency selective periodic structures play a vital role in medical telemetry. Antennas are either implanted inside human bodies or mounted on the clothes to establish a communication link between the medical devices kept with the patient and remote medical units or hospitals within short range, leading to a continuous monitoring system [3, 7]. However, antennas when kept with the patient may cause unwanted radiation towards the body while transmitting data. This may also have harmful effects over long durations [8, 9]. FSSs with reflective natures, being high impedance structures, can be used to minimize this unwanted radiation by integrating with the antenna. Metamaterials, with their unusual properties, exhibit several distinct features while implemented in antennas and FSSs, such as miniaturization of the antenna profile, gain enhancement of antennas, reduction in back radiation using FSSs, etc [10–12]. Thus, metamaterial-based antennas and FSS structures are significantly useful in biomedical devices for body implantable or non-implantable units, detailed discussion on which is provided in this chapter.

7.1 Introduction

Biomedical engineering has gained significant attention from researchers as well as industry since past few years and the trend is growing at a high rate because of the developments in healthcare [1, 5]. Apart from academia and research this has a prominent role in improving medical diagnosis and treatment. The use of biomedical engineering in medical diagnosis and cure involves monitoring various parameters of human body including temperature, pulse rate, blood pressure, heart rate, glucose level, BMI, etc [13, 14]. Continuous monitoring of these parameters has become essential for critical and old patients. However, visiting hospitals and clinics can be difficult for such patients for various reasons and this problem can be overcome with the help of biomedical telemetry. With the development of the healthcare industry, biomedical telemetry has emerged and made it practicable to monitor various health symptoms of a patient remotely, minimizing frequent medical check-ups and hospital appointments [15]. Biomedical telemetry is mostly implemented in the ISM bands in between 900 MHz and 5.8 GHz and in some cases beyond the range [16, 17]. Apart from monitoring health parameters mentioned above, presently complicated procedures such as insulin pump therapy, endoscopy, and even deep brain simulations can take the advantage of remote monitoring systems [18, 19]. Monitoring certain health parameters may need to place in sensors or measuring equipments inside human body wholly or partially in an invasive manner known as implantable unit. Body implantable sensors or antennas are useful for applications such as monitoring blood pressure and temperature, diagnosis of heart diseases, tracking dependent old people, etc [17]. Diagnostic information from an electronic device implanted inside the human body (such as a pacemaker) is transmitted to an external RF receiver. This communication is referred to as body-centric wireless communication or simply body-centric communication [20].

Body-centric communication has become an essential part of our daily life and it is presently a immensely growing field of body-centric wireless networks (BCWNs) simply known as body area networks (BAN). The whole body area network is demonstrated in figure 7.1 [21]. It consists of various stages, such as: the implantable or non-implantable units composed of sensor antennas attached to the human body from which the information (health parameters as discussed) goes to the personal digital assistant (PDA) module, which is most commonly a smart phone [22]. The information is then transmitted to the access point or base station from where, via the internet, the information reaches hospital doctors or medical researchers for further diagnosis. The assistance and treatment as required by the patient is provided by the hospital following this process on the reverse path in the communication network. Body-centric wireless communication can be of primarily three types: on-body, off-body and in-body communications, as classified below [20, 23].

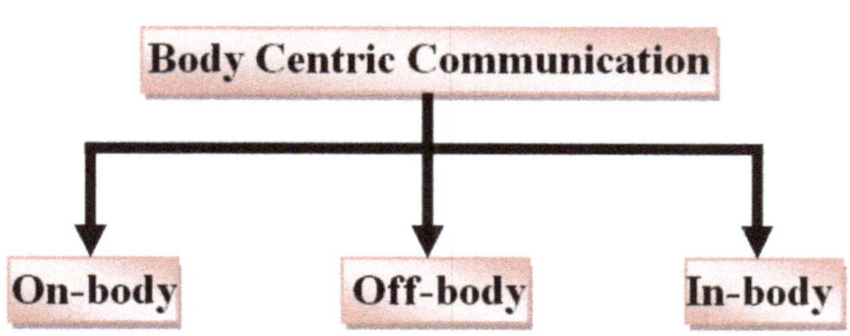

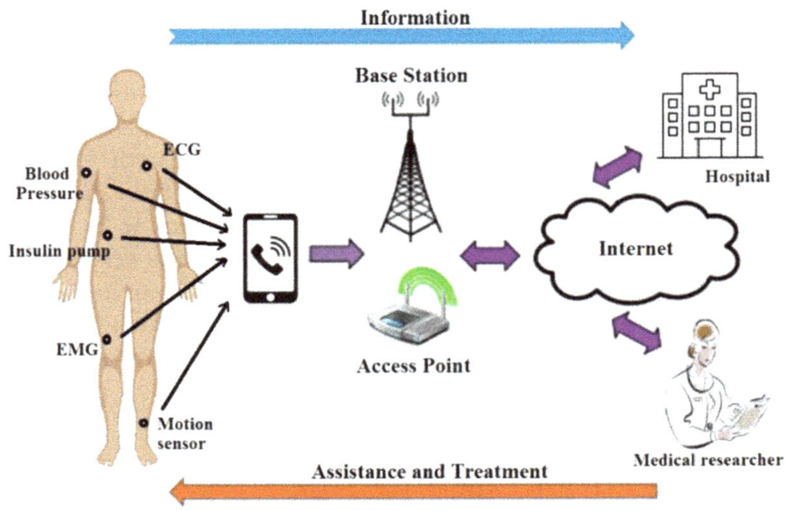

Figure 7.1. Body-centric communication in a body area network. Reprinted by permission from Springer Nature Customer Service Centre GmbH: [Nature][Advances in Mobile and Wireless Communications] [21], Copyright (2008)

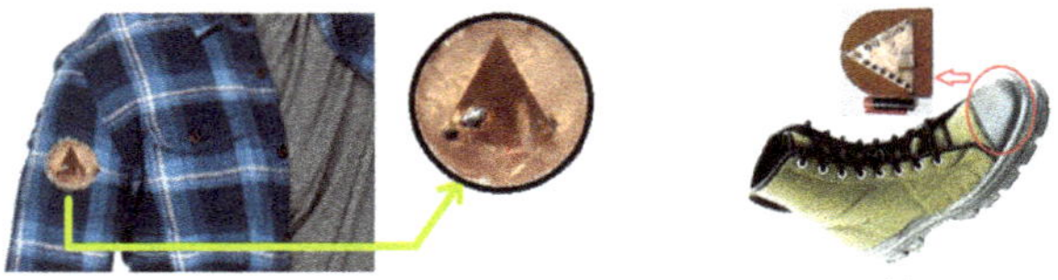

Figure 7.2. Wearable antennas for on-body biomedical application; (a) antenna mounted on shirt. [24] John Wiley & Sons. [© 2015 Wiley Periodicals, Inc.] (b) antenna mounted on the shoe. [25] John Wiley & Sons. [© 2020 The Institution of Engineering and Technology]

Wearable devices exhibit on-body communication. The communication between sensor modules mounted over the skin, such as temperature sensors, ECG sensors, motion sensors, etc., and the antenna module attached to the clothes is referred as on-body communication [22]. Two wearable applications involving on-body antennas are shown in figure 7.2, one of which is mounted on the shirt worn by a person (left) [24] whereas the other is supposed to be mounted on the front end of the shoe (right) [25]. The communication involving an on-body device such as a sensor or antenna module and an external device such as a smart phone or local access point is referred as off-body communication. In-body communication refers to the communication between an implantable module placed inside the body through invasion and an external device such as a smart phone or Bluetooth device attached to the clothes [23].

Body-worn sensors and antennas are much more feasible for biomedical telemetry than implantable sensors and antennas. Implantable modules, more popularly known as implantable medical devices (IMDs), are made up of components including battery, antenna, and sensors and thus various challenges take place in designing such modules, such as miniaturization, compatibility with the human body and the patient safety [5, 7]. Lossy tissues with less electromagnetic compatibility inside the human body add to the design complexity of the IMDs.

The antenna is the primary building block for an implantable module as the basic requirement of signal transmission from body to external access point or remote centre depends on the antenna working. Miniaturization of the antenna is required by implantable or non-implantable modules [24, 26, 27], however, low-profile antennas often exhibit poor gain that can be negative gain at times [28]. Thus, there is a trade-off between miniaturization of antenna dimensions and adequate antenna gain, both of which are difficult to accomplish simultaneously. This problem can be overcome by implementing metamaterials in the design. Metamaterials possess unusual properties, such as negative permittivity ($\varepsilon < 0$) or negative permeability ($\mu < 0$) or both [29, 30]. As a result, metamaterials exhibit distinct features that make them suitable for various applications, such as absorbers, cloaking, radar cross section reduction, miniaturization of the dimensions, enhancement of antenna radiation, etc [11, 12, 27]. The use of metamaterial-based antennas is a good choice for biomedical applications (on-body and in-body).

Another challenge in designing antennas for biomedical applications involves back radiation where the antenna mounted on the body or implanted inside causes unwanted radiation towards the human body, which may have hazardous effects over long-term use [31–33]. Such a problem can be overcome with the use of frequency selective surfaces (FSSs). FSSs have a periodic structure in the form of an array of metallic patches or apertures based on a dielectric substrate [34], among which patch-type FSSs are reflective in nature along with high impedance properties that make them suitable for reducing back radiation from antennas (especially bidirectional antennas) [35]. This can protect the patient using on-body antenna modules from unwanted microwave radiation. Moreover, the use of FSSs placed below the antenna as a substrate or above the antenna as a superstrate can help in enhancing antenna gain in the broadside direction [27]. The use of metamaterial-based FSSs can elevate these advantages, because metamaterial-based periodic structures have sub-wavelength dimensions [36] on the order of $\lambda/10$ or less and thus overall dimension of the wearable module remains low in profile along with the ability to use a large number of unit cells.

7.2 Metamaterial based antennas for biomedical applications

The use of metamaterials in designing antennas for biomedical applications incorporate various classes of metamaterials leading to various metamaterial-based antennas such as complementary split ring resonator (CSRR) loaded antenna, composite right/left handed (CRLH) transmission line (TL) based antenna, zero-order resonance (ZOR) antenna, antenna with metasurface etc. Metamaterial unit cells, metamaterial-inspired structures can be implemented in antennas for both on-body and in-body applications. In this section design method, characteristics and distinct features of both implantable as well as wearable antennas using metamaterial loading are discussed.

7.2.1 MTM-based on-body antennas

As discussed earlier, the design of wearable antennas for on-body biomedical applications is more feasible compared to the implantable antennas. Metamaterials can be used to design such antennas in several ways. The radiating patch of the antenna can be loaded with MTM structures such as CSRR, CRLH, etc., [37–41] whereas structures with a periodic array of MTM-based unit cells such as electromagnetic band gap (EBG) structures, frequency selective surfaces (FSSs), artificial magnetic conductors (AMCs) and metasurfaces can be attached to the antenna as a substrate or superstrate in order to realize MTM-based wearable antennas [27, 35]. Flexible and ultrathin dielectric materials such as polymer, nanomaterial, polyethylene, etc., should be chosen for on-body antennas [42, 43] so as to mount them on wearable clothes without degrading antenna performance. Textile materials such as cotton, felt, denim etc., are also frequently used as substrates for on-body antennas [44]. However, in such cases the relative permittivity of the material should either be known or be tested before. Regarding characteristic analysis for wearable antennas, certain parameters should be studied necessarily, such as front-to-back ratio (F/B) and specific absorption rate (SAR) for 1 g or 10 g tissue [45].

7.2.1.1 MTM loaded on the antenna patch

In most of the MTM-based on-body antennas metamaterial-based structures are loaded on the antenna patch. This is incorporated in the patch either with a unit cell of MTM or multiple MTM cells. The use of zeroth-order mode in a metamaterial-based compact antenna with reconfigurable radiation pattern at 2.4 GHz was proposed by Yan and Vandenbosch [37]. The radiating patch, as shown in figure 7.3 (a), is made of conductive textile with $\sigma = 1.18 \times 10^5$ S m^{-1} above a felt substrate with $\varepsilon_r = 1.3$ and loss tangent 0.044. The patch consists of vias on the edges connected via RF switching diodes to the patch. The antenna operates as a conventional microstrip patch with the switches off (vias disconnected) and exhibits broadside radiation in the first-order mode ($n = 1$), as shown in figure 7.3(d) at 2.4 GHz (right). At the same frequency with the switches on the antenna operates in the zeroth-order mode (ZOR) and radiates like a monopole antenna as shown (left). The antenna exhibits gain up to 3.9 dB, whereas the radiation efficiency remains low up to 45% at monopole mode. The z component of electric field distribution depicted in figure 7.3(b) also justifies the patch mode for switch off ($n = 1$) and monopole mode for switch on ($n = 0$).

The authors have investigated compatibility of the antenna for on-body applications by placing it on a human tissue model made of skin, fat, and muscle layers as shown in figure 7.3(c) under bending conditions. The antenna operates well on the bent tissue model without significant change in the resonance or degradation in radiation. The SAR values for the antenna are also well below the standards, such as

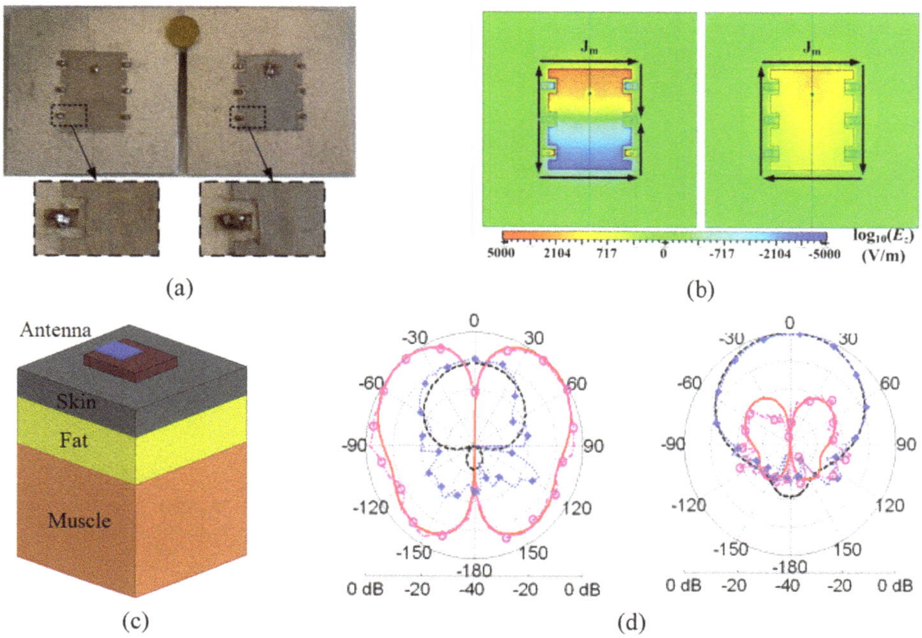

Figure 7.3. (a) Fabricated MTM-loaded patch antenna, (b) electric field distribution at 2.45 GHz, (c) human tissue mode with the layers, and (d) reconfigurable radiations at ZOR and +1 mode. Reproduced from [37].

0.05 and 0.01 W kg^{-1} for the ZOR mode and the +1 mode, respectively, while averaged over 10 g of biological tissue.

The use of metamaterial-based unit cell on the patch of a printed monopole antenna was proposed by Roy and Chakraborty [38] for on-body application over two wide frequency bands of 1.6–2.56 GHz and 4.24–7 GHz including WiMAX, GSM 1800 MHz as well as IEEE 802.11 a/b/g/n WLAN operating frequencies. The metamaterial unit cell incorporated on the patch, was first analyzed for the MTM properties on a 1 mm thick jeans-based substrate with dielectric constant of 1.6, determined using VNA and a dielectric assessment kit. The MTM cell used was found to possess negative permittivity and negative permeability in different frequency bands within the desired operating range.

The antenna with MTM unit cell on the patch as proposed is shown in figure 7.4 (a). It was shown that with thicker substrate (jeans) the upper band was shifted to the right covering both the WLAN bands. With the use of MTMs, antenna dimension is reduced to $0.1\lambda \times 0.1\lambda$, whereas antenna gain reaches 1.9 dB at 2.4 GHz and 5.9 dB at 5.8 GHz. It was realized that under bending condition with the bending radius between 30 and 205 mm, field pattern of the antenna got changed leading to variation in the antenna operating bandwidth. In order to perform human body analysis of the antenna a human tissue model was developed by the authors as shown in figure 7.4(b) with the material properties as provided in table 7.1, including relative permittivity and conductivity of the various layers beneath the skin. The

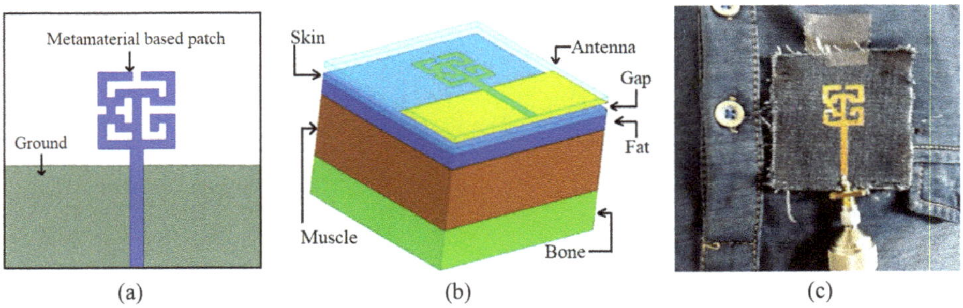

Figure 7.4. (a) Monopole antenna with MTM-based unit cell on the patch, (b) human body model made of skin, fat, muscle and bone, and (c) antenna designed on jeans substrate mounted on the shirt of a person. Reprinted by permission from Springer Nature Customer Service Centre GmbH: [Nature][Wireless Personal Communications] [38], Copyright (2019).

Table 7.1. Material properties of various layers of the human tissue model. Reprinted by permission from Springer Nature Customer Service Centre GmbH: [Nature][Wireless Personal Communications] [38], Copyright (2019).

Properties	Skin	Fat	Muscle	Bone
Permittivity	37.95	5.27	52.67	18.49
Conductivity (S m^{-1})	1.49	0.11	1.77	0.82

antenna fabricated on jeans was tested on various body parts including arms, legs, chest and the performance was more satisfactory while placed on the chest, as shown in figure 7.4(c).

In the above design the unit cell of the MTM structure was used; however, the unusual properties of metamaterials can be best explored while used as an array of unit cells [39–41]. This is because in the array structure inductors and capacitors between the consecutive unit cells can be realized that is not realizable when the unit cell is used alone on the antenna patch. This was demonstrated by AlSabbagh *et al* [39] by incorporating an array of 3 × 5 Hilbert-shaped metamaterial-inspired unit cells on the radiating patch of the antenna. The MTM-inspired antenna was intended to exhibit triple-band response for wearable devices operating in the Long-Term Evolution (LTE) and Long-Term Evolution-Advanced (LTE-A) frequencies used by mobile communication as well as Wireless Local Area Network (WLAN), and Worldwide Interoperability for Microwave Access (WiMAX) bands. The radiation pattern of the antenna is also versatile including steerable radiation in middle band and directive end-fire radiation in the upper band. The radiation in the lower band is suitable for short-range communication.

The patch, as shown in figure 7.5(a), consists of a MTM EBG structure with anisotropic band gap realized using third-order Hilbert curve along with a tuning T-shaped stub. Inkjet printing was used to fabricate the antenna, as shown in figure 7.5(c). With the step-by-step design process the authors showed that with the dominant

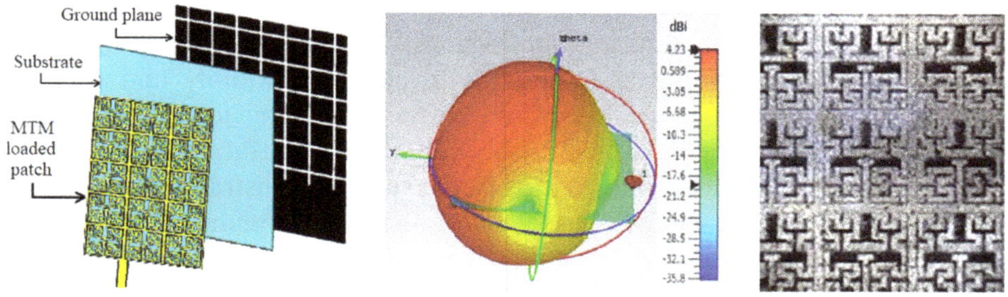

(a) (b) (c)

Figure 7.5. (a) Antenna patch loaded with MTM-inspired EBG array, (b) end-fire radiation from the antenna at 5.8 GHz, and (c) fabricated prototype of the antenna with the MTM cells enlarged. [39] John Wiley & Sons. [© 2019 Wiley Periodicals, Inc.]

(a) (b) (c) (d)

Figure 7.6. (a) CPW-fed antenna with MTM-based complementary cells loaded on the patch (b) fabricated antenna (c) voxel model of human body (d) antenna mounted on a cylindrical structure. Reproduced from [40]. CC BY 4.0.

current distribution on the MTM-loaded patch that mimics the geometry, an end-fire radiator pattern (along x) is achieved at 5.8 GHz as shown in figure 7.5(b). It was also realized that with the use of partial ground plane and defects on it (square EBG) relatively wider bandwidths can be achieved in the three frequency bands. Moreover, the radiation is steerable between 0° and 80° in the 3.3–3.9 GHz band. The antenna exhibits positive gain in the 3.5 and 5.8 GHz bands, whereas the gain is about −5 dBi at 2.45 GHz, which makes it suitable for short-range communication between a wearable sensor and handheld devices or nearby base station. Characteristics of the antenna prototype made using silver nanoparticles do not vary significantly from the copper-made patch. The antenna characteristics were shown to be satisfactory while measured on a phantom model mounted on the helmet.

More distinct advantages of incorporation of an MTM array on the antenna patch can be found in the work proposed by Al-Adhami and Ercelebi [40], including lowering of specific absorption rate (SAR) value. The design can be visualized from figure 7.6(a). In the design the authors have included a complementary Minkowski fractal MTM array on a conventional CPW-fed monopole antenna patch.

It resonates at several ISM bands including 403, 433, 611, 912 MHz, and 2.45 GHz for remote health-monitoring applications. The antenna was designed on a 0.3 mm thick low-loss polyamide substrate to achieve flexibility and robustness (figure 7.6(b)). Two separate stub extensions are included in the feed line that besides improving the impedance matching also suppress the surface waves along the antenna edges. Further reflection of surface wave from substrate edges are minimized by etching out similar MTM slots from the ground plane. In this case, the authors also fabricated the antenna using silver nanoparticles with a conductivity of 1.3×10^6 S m^{-1}. It was observed that with the increase in MTM array size antenna, bandwidth got enhanced significantly at most of the operating bands and an average gain of 2 dBi is retained. The antenna was studied at bending angles up to 45° (figure 7.6(d)) and when mounted on the voxel model of human body (figure 7.6(c)), it showed satisfactory performance. Moreover, the electric field leakage towards the human body was kept between 100–133 mV m^{-1}. Minkowski geometry was used to dominate the magnetic field more than electric field, leading to more control on SAR as the human body permeability is unity. The head SAR was maintained within a safe limit between 0.32 and 0.54 W kg^{-1}.

7.2.1.2 MTM periodic structures as substrate

We have seen how MTM-based array structures can be used on the antenna patch or ground plane. However, this increases the design complexity. Moreover, for antennas with bidirectional radiation, such designs would not reduce the back radiation towards the human body. MTM-based periodic structures can be placed as a substrate or reflector below the radiating element in the wearable antenna at much smaller spacing compared to the wavelength [46–49]. Ashyap *et al* incorporated MTMs on a U-shaped printed monopole antenna in the form of an artificial magnetic conductor (AMC) at 2.4 GHz [46]. The primary objective was to reduce the effect of frequency detuning and to reduce the back radiation especially when mounted on the human body. Jeans fabric is used as a substrate with the loss tangent and permittivity of 0.05–1.68, respectively, whereas Shieldit Super is used as the conductive material with a conductivity of 1.8×10^5 S m^{-1} for the proposed U-shaped monopole antenna as shown in figure 7.7(a).

The AMC unit cell is composed of a metallic split ring resonator (SRR) surrounded by a square loop with a ground plane below, as shown in figure 7.7 (b). It was observed that the MTM-based AMC has an in-phase response between −180° and + 180° with 0° at 2.4 GHz. This led to improvement in the impedance matching of the antenna while integrated together. With the AMC the antenna gain enhanced from 2 up to 7.5 dBi, whereas with PEC the enhancement was only up to 5.3 dBi. Moreover, the front-to-back ratio (FBR) was significantly increased by up to 13.7 dB leading to low FBR, which is essential for wearable applications to reduce the radiation effects on the human body. This in turn reduced the SAR level of the antenna while mounted on different human body parts such as the chest and arm. Interestingly it was evident that without MTM, the antenna resonance shifted from desired 2.4 GHz whereas with the MTM-based AMC the shift was not significant. This occurred because of the isolation provided by the AMC between antenna and

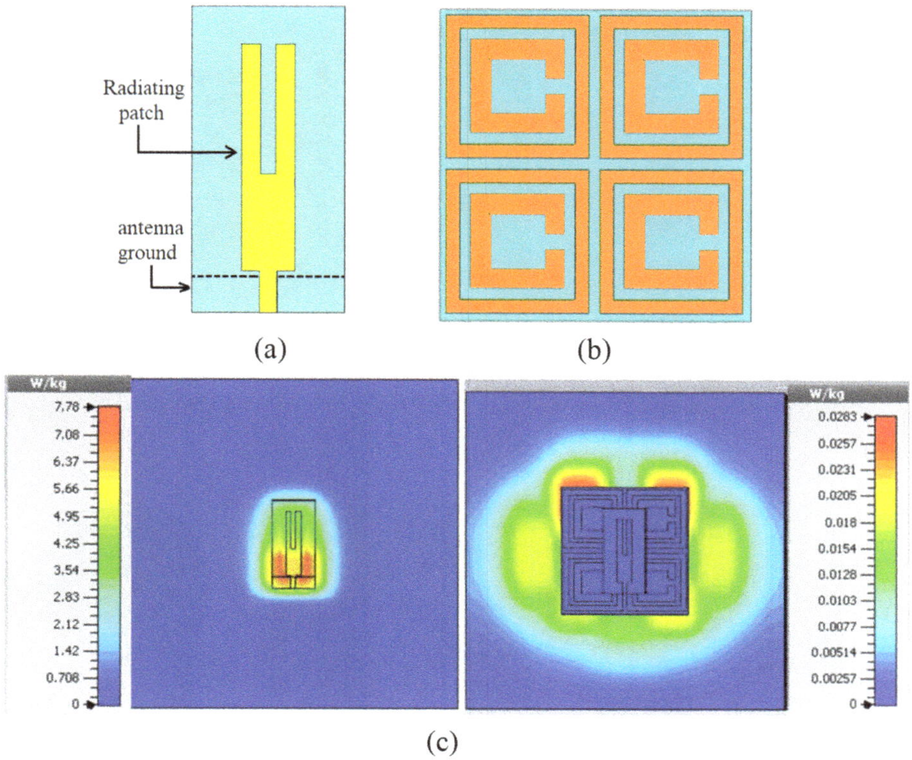

Figure 7.7. (a) U-shaped monopole antenna (b) MTM-based artificial magnetic conductor (AMC) (c) SAR for antenna with (right) and without AMC (left). Reproduced from [46]. CC B Y 3.0.

body tissues. The SAR level of the antenna was observed to be reduced by a large amount from 7.78 to 0.028 W kg^{-1} while in use along with the MTM-based AMC for on-body application, as evident from figure 7.7(c).

The use of embroidery in designing wearable antennas is of utmost importance as it increases the degree of material wearability in comparison to the conventional antennas with metallic patches. Such technology also makes the antenna more compatible with various body parts with different bending angles such as the chest (lower bending), arm (more bending), etc.

Gil *et al* proposed [47] an embroidered MTM-based monopole antenna at 2.45 GHz where a split ring resonator (SRR)-based EBG structure is used as a substrate at a minimal spacing from the antenna, as shown in figure 7.8(a). The antenna is constructed on a three-layered felt structure among which the monopole in the top and SRR-based EBG with monopole ground in the middle are made of Amberstrand Silver 66 yarn, whereas the EBG ground plane made of 0.2 mm thick Nora Dell is included in the bottom layer. With the use of MTMs the antenna acquires a gain of 7.8 dBi, whereas the front-to-back ratio enhanced by 10 dB and these are comparable to the previous deign. In order to investigate the compatibility of the antenna for wearable biomedical applications it was mounted on a

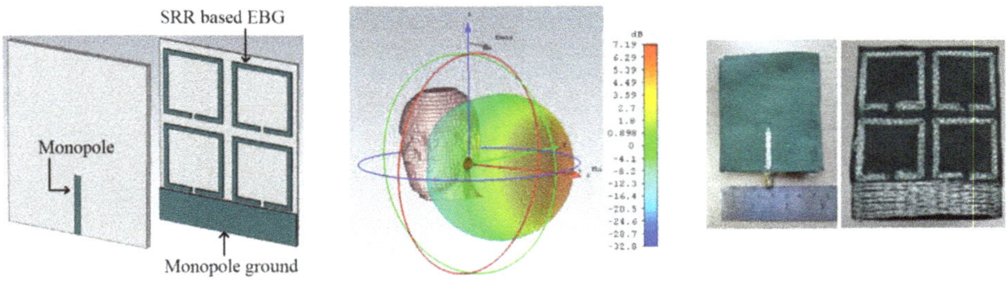

Figure 7.8. (a) Monopole antenna with MTM-based EBG as substrate (b) radiation from the antenna with MTM located at ear of the Gustav model (c) Fabricated antenna and MTM using embroidery on felt. [47] John Wiley & Sons. [© 2018 WILEY-VCH Verlag GmbH & Co. KGaA, Weinheim]

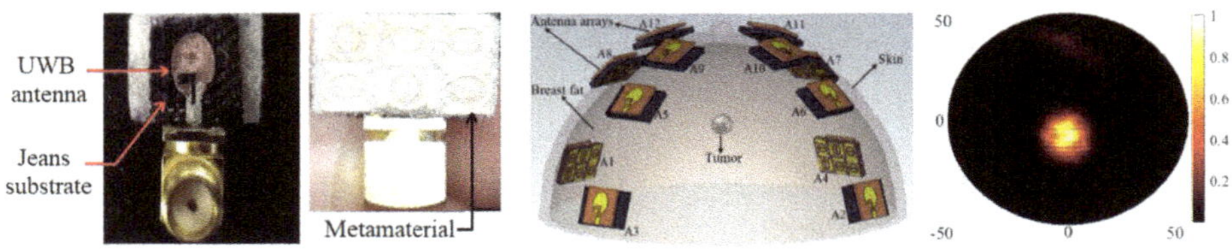

Figure 7.9. (a) MTM-based UWB antenna on flexible substrates (MTM shown on right) (b) breast model with 12 antenna-MTM composite arrays for tumor detection (c) image reconstructed with tumor. Reprinted from [48], with the permission of AIP Publishing.

heterogeneous voxel model Gustav based on 57 biological tissues. The lossy medium reduced antenna gain and efficiency to −2.44 dBi and 13.5%, respectively. However, with MTM the gain and efficiency both were enhanced to 7.5 dBi and 70%. The importance of metamaterials in wearable or on-body antennas is justified and evident from this. The monopole MTM exhibits a much reduced SAR value, up to 0.119 W kg^{-1} for 10 g tissue.

The application of metamaterial-based array structures in ultra wideband (UWB) antennas for biomedical applications is a challenging job. Alhawari *et al* proposed a MTM-based flexible UWB antenna [48] for breast imaging and wireless body area network (WBAN) applications.

Breast imaging has enormous significance in medical diagnosis, including tumor detection and early detection of breast cancer and various other medical issues. An SRR-based unit cell was used to form the MTM on a textile felt substrate ($\varepsilon_r = 1.4$) that was placed below the elliptical patch antenna based on a denim substrate ($\varepsilon_r = 1.2$), as shown in figure 7.9(a). It was observed from the EM simulation that initially the UWB antenna had some stop bands or notches in the operating band and the SRR-based MTM substrate eliminated them. The antenna exhibits a peak gain of 8.85 dBi and a maximum radiation efficiency of 94% at the frequency of 30 GHz with the inclusion of the MTM substrate. Higher gain makes the antenna suitable for medical imaging applications. The antenna functioned well with a good reflection coefficient (S11) response when placed on the human body. An array of 12 such MTM antennas was mounted on the breast model with a tumor sample inside (figure 7.9(c)). One UWB antenna was used to transmit the pulse whereas other sensors were intended to receive the signal. All the received signals were extracted from a CST MWS EM simulator and imported to MATLAB where the image was reconstructed using a time-reversal algorithm. It was evident that the tumor sample of 4 mm diameter inside the breast model of up to 150 mm diameter was successfully detected in the image. However, the accuracy increased with the number of antenna-MTM arrays around the tumor. Also, the SAR level was far below the standard 2 W kg^{-1} for 1 g tissue.

7.2.2 MTM-based implantable antennas

Implantable antennas are an essential part of implantable medical devices (IMDs) and stimulators such as implanted sensors, medicine infusion devices, nerve stimulators, cochlear implants, defibrillators, and leadless pacemakers [50–52]. IMDs have been of utmost significance for patients, especially old age and critically ill patients, who cannot visit hospital frequently. However, there are several requirements for designing implantable antennas including low profile, adequate gain, biocompatibility, high level of flexibility and low specific absorption rate (SAR) [14, 16, 17]. The asymmetric antagonistic environment inside a human body makes implantable antenna design more challenging. Implantable antennas cannot work alone and need to be integrated with additional microelectronic components such as sensors, circuitry, and batteries [16, 53]. Thus, compatibility of the antenna with these components is essential. Metamaterials can be incorporated in implantable antennas to achieve miniaturization of dimension, reduction in back radiation and SAR, enhancement of radiation in the desired broadside direction [54–57]. MTMs can be integrated with the radiating element or can be placed as a superstrate above the antenna.

Bhattacharjee et al. proposed an implantable CPW fed antenna with an MTM-inspired radiating element [54] for biomedical applications in the ISM band frequency of 1.35–3.5 GHz. Asymmetric complementary split-ring resonators have been used by the authors through multiple mode excitations in developing the radiating element. The antenna patch is based on a silicon substrate with ε_r of 11.7 and 275 μm thickness and the radiating element is built using 400 nm thick gold. These materials were chosen due to their biocompatibility with the inside of human body. The CSRR used in the patch has different widths for rings and is composed of asymmetric arc-shaped slots as shown in figure 7.10(a). Characteristic modal analysis is implemented for bandwidth enhancement and miniaturization. In the choice of phantom both single layer and multi-layer phantoms are considered for the antenna implementation, where the Cole-Cole model proposed dielectric parameter has been utilized as shown in figure 7.10(b).

The antenna was placed at a depth of 1 mm from the skin layer during simulation using CST and in HFSS simulation; it is placed at a depth of 13.2 mm from the top

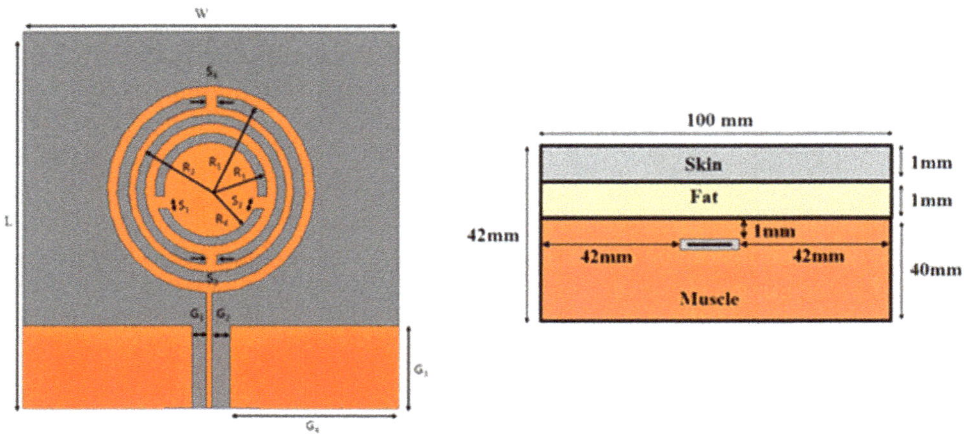

(a) (b)

Figure 7.10. (a) Implantable CSRR loaded antenna (b) CST multi-layered tissue phantom model [54]. John Wiley & Sons. [© 2018 The Institution of Engineering and Technology]

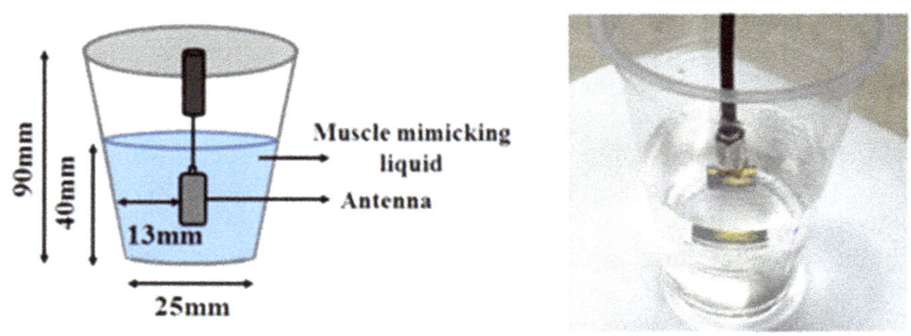

(a) (b)

Figure 7.11. (a) The antenna placed inside cup (b) Measurement inside muscle mimicking liquid [54] John Wiley & Sons. [© 2018 The Institution of Engineering and Technology].

of muscle equivalent phantom. During fabrication the oxide layer thickness was varied from 700 to 1000 nm for radiation in the ISM band. The measurement setup is shown in the figure 7.11. It was observed that the antenna S11 response shifts left when placed inside muscle mimicking liquid. The wideband response of the antenna makes it robust to detuning effects which are caused due to dynamic permittivity values and dispersive properties of different human tissues. The antenna offers a peak gain of -16.5 dBi over the ISM band (2.4–2.48 GHz).

The implantable antenna-MTM structure discussed above works on linear polarization. However, the antenna, once implanted inside the body, cannot be relocated or rotated easily and with frequent change is body posture the antenna may transmit/receive signals of different polarization. To overcome this problem Zada *et al* came up with a dual-band circularly polarized antenna [55] loaded with a metamaterial-based superstrate, as shown in figure 7.11(a), implantable inside various parts of the human body including the heart and brain. The antenna patch has a serpentine shape with a ground plane below. A 0.127 mm thick Rogers RT/duriod 6010 dielectric with permittivity of 10.2 and loss tangent of 0.0035 is chosen for both the antenna and MTM superstrate. High permittivity dielectric isolates the antenna from the lossy surroundings inside human body and smaller loss tangent retains the desired antenna radiation without significant degradation. An array of MTM cells are placed above the antenna as superstrate. The authors investigated the antenna along with the necessary microelectronic components inside a biocompatible ceramic alumina (Al_2O_3) container using HFSS EM simulator. The antenna with MTM was implanted inside the body of a human model [58] at various locations including scalp, heart, stomach, and intestines, as shown in figure 7.11(b). The MTM, because of its EVL property, also led to a very low profile of the antenna system with the dimensions 7 mm × 6 mm × 0.254 mm.

It was observed that the antenna-MTM inside the human body exhibits resonances at 915 and 2450 MHz with adequate impedance matching and impedance bandwidth of 17.8%–35.8%, respectively. The frequency shift was observed to be negligible even in such a heterogeneous environment because of the MTM. Interestingly, the 3 dB axial-ratio (AR) bandwidth in both the ISM bands got increased with the MTM to 21%–17% in the lower and upper bands. An average gain enhancement of 1.5 dB was evident with the use of MTM superstrate with peak gain values of −17.1 and −9.81 dBi. With the metamaterial array the antenna exhibits omni-directional radiation, with maximum radiation outwards from the model, which is required for implantable applications. The MTM also ensures significant reduction in the SAR level from 576 to 394 W kg^{-1} over 1-g of tissue for human scalp.

7.3 Metamaterial based FSSs for biomedical applications

This section emphasizes on the implementation of MTM-based FSSs and similar periodic structures for various biomedical applications. In the previous section in some cases the use of MTM-based array structures for wearable on-body and in-body implantable antennas were observed for antenna miniaturization, gain augmentation and SAR reduction, etc [54–57]. However, the field of biomedical engineering has many other distinct applications where MTM-based FSS structures are found to be in use and such areas are focused on in this chapter. These include biosensors, wireless power transfer (WPT) control in biomedical devices, wearable vital-sign sensing, and medical imaging [59–62]. Various MTM-based frequency selective structures including FSS, EBG, AMC and metasurfaces are found to be used for such applications with unit cell dimensions much smaller compared to the traditional half-wavelength or quarter-wavelength value ($<<\lambda/4$) and are designed on ultrathin flexible dielectric substrates.

7.3.1 Biomedical sensing

Biomedical sensors are electronic devices or transducers that are used to convert biomedical signals into measurable electric signals. Biomedical sensors are found to be of great interest in the field of medicine, microbiology, medical diagnosis and treatment [59–61]. Sensors suitable for biomedical applications are used for measuring molecular concentrations in biological tissues, vital-sign monitoring, pH level estimation, investigating DNA, etc. An electromagnetic signal impinges on a material under test (biological tissue) that is placed close to a MTM-based array structure, such as a FSS, followed by a detector to investigate the signal after the array [61, 63], as shown in figure 7.12. The variation in the electrical properties caused by the presence of the material under test leads to a variation of the MTM-based sensor resonant frequency. This shift in resonance frequency helps in medical diagnosis such as tumor detection.

Emami-Nejad and Mir proposed [61] the design of a split ring resonator (SRR)-based FSS for cancerous cell detection in a biological tissue. The authors have used SRR-based metamaterials in the proposed biosensor because of the fact that SRRs and CSRRs are highly sensitive to any change in the characteristic electromagnetic waves. The unit cell of the FSS contains a SRR slot with a split gap of 1 mm, leading to a bandpass response at the resonating frequency without any sample placed on the biosensor.

The biosensor structure with an array of SRR slots on a screen of thin layer of gold, as shown in the figure 7.11(a), is based on a Teflon substrate with a permittivity of 2.1. The structure resonates at 4.6 GHz without any sample placed on it, whereas with a normal cell placed on the SRR the frequency was observed to be shifted to 3.35 GHz. It was observed in the studies that with a malignant cancer cell placed on the SRR, resonance was shifted to 3.1 GHz, whereas with a cirrhotic cancer cell the shift occurred to 3.2 GHz. Small variations in resonance against different types of cancer cells is the primary disadvantage of the biosensor proposed by the authors. However, it was also observed that with the increase in split gap (up to 2 mm) in the SRR-based biosensor the amount of frequency shift due to change in cancer cells also increased up to 0.75 GHz. The effect of thickness of the sample on the detection was also investigated, which shows that with an increased sample thickness the resonance frequency got shifted to the higher end, which can lead to good resolution of the biosensor. The electric field distribution on the SRR slot was observed with and without a sample on the biosensor and as can be seen from figure 7.13 (b) and (c), the field strength varies significantly with the use of the sample.

The application of metamaterial-inspired biosensors can be extended towards detection of fatal diseases like cancer at an early stage [62, 64, 65]. Metamaterial-based

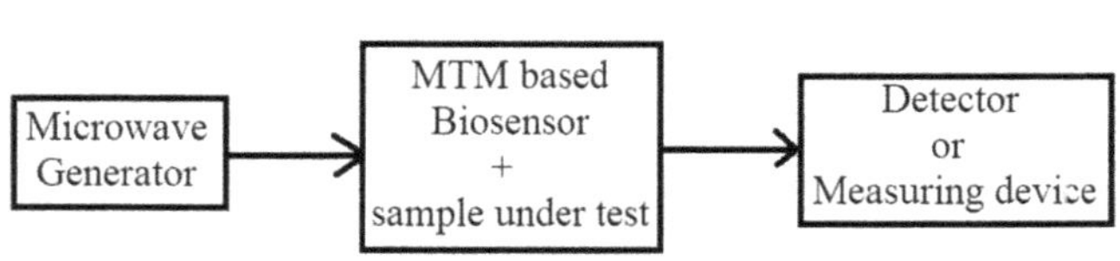

Figure 7.12. A simple block diagram of a MTM-based biomedical sensor.

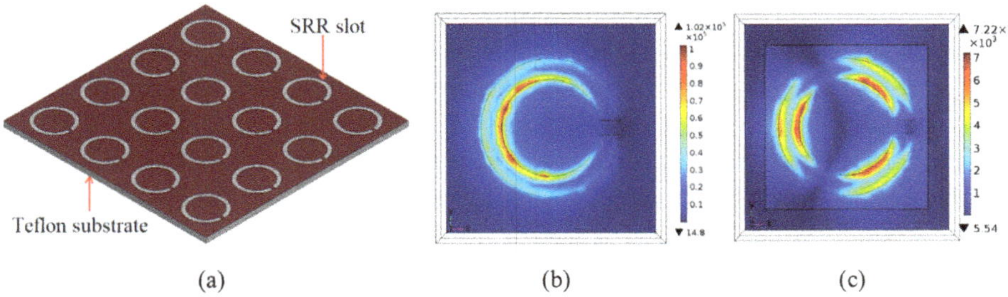

Figure 7.13. (a) SRR slot-based FSS as biosensor (b) electric field distribution on the SRR when no sample is placed (c) electric field distribution on the SRR when sample with cancer cells are placed. Reprinted by permission from Springer Nature Customer Service Centre GmbH: [Nature][Optical and Quantum Electronics] [61], Copyright (2017)

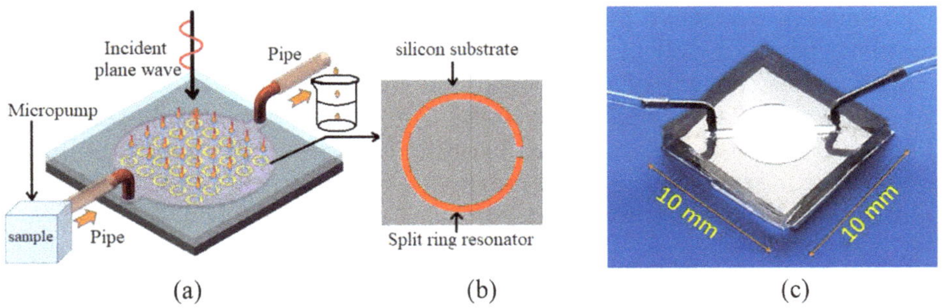

Figure 7.14. (a) SRR-based biosensor chip integrated with microfluidics (b) SRR unit cell with single split (c) fabricated biosensor with PDMS microfluidic on the microchannel. Reproduced [66]. CC BY 4.0.

periodic structures such as FSSs operating in the THz frequency regime are significantly useful for such applications because of their precisely and finely tunable resonating frequency, which is compatible with the frequency of vibration of certain cancer biomarkers. According to the National Cancer Institute, a biomarker is 'a biological molecule found in blood, other body fluids, or tissues that is a sign of a normal or abnormal process, or of a condition or disease,' such as cancer [63]. Cancer biomarkers include proteins, RNA, and DNA. With the advancement of micro-nanofabrication technology, metamaterials at THz became sensitive to variation in micro-environment on the metamaterial surface. Asymmetric metamaterial arrays at THz exhibit Fano resonance, which is extremely useful in detecting biomolecules with lower concentration. However, THz metamaterials possess a high amount of water absorption, which limits its use to the diagnosis of dry specimen.

The challenges were overcome by Geng *et al* [66] with the use of microfluidics in THz metamaterial structures due to its little usage of liquid and tight fluidic confinement. They have demonstrated the implementation of an SRR-based MTM array integrated with microfluidics to diagnose the biomarker of liver cancer at early stage. The arrangement is shown in figure 7.14(a). Two types of SRR-based metamaterial unit cells are proposed with single-split and two-split gaps, the first of which is shown in figure 7.14(b). The SRR is designed here on a silicon substrate with high resistivity. A small amount of the sample enabled by the microfluidics is flowed through a microchannel by using the microelectromechanical (MEMS) technology. Biocompatible polydimethylsiloxane (PDMS) was used to fabricate the channel in the biosensor chip. The biosensor chip integrated with PDMS is shown in figure 7.14(c). With the variation in the cancer biomarker sample placed above the SRR-based MTM, capacitance of the split gap changes, which in turn causes sharp changes in the resonating frequency of the incident plane wave. This frequency variation is responsible for sensing the biomarker.

Liver cancer biomarkers such as alpha fetoprotein (AFP) and glutamine transferase isozymes II (GGT-II) antigen were placed above the SRR—MTM array based on a 4 inch high resistivity silicon wafer. It was observed by the authors from detailed analysis using the Finite Difference Time Domain (FDTD) method as well as experimental verification that the resonance frequency reduces with an increase in the surrounding refractive index. It was shown that any variation in the liquid permittivity causes a shift of resonance to lower frequency. The authors observed that for a smaller split gap of 2 μm a maximum frequency shift of 140 GHz $R^{-1}IU^{-1}$ can be achieved (RIU = Refractive Index Unit). Sensitivity and Q factor of the MTM-based biosensor was increased by using two splits (120° apart) on the SRR due to the occurrence of two Fano resonances. Frequency shift of 140 and 150 GHz/RIU were observed at the two resonances, leading to high sensitivity.

7.3.2 Wireless power transmission enhancement

The design of implantable biomedical devices (IMDs) are challenging due to various factors discussed in section 4.2.2. However, for certain applications such as monitoring vital signs (pulse rate, blood oxygen, blood glucose level, etc), remote drug delivery, insulin pumping, hyperthermia for cancer treatment, cardiac pacemakers, etc., implantable devices integrated with antennas have become essential [67–69]. The development of implantable devices has certain primary requirements such as lower weight, biocompatibility, longer battery lifetime and reliability, cost effectiveness, etc. In spite of longer battery life certain IMDs with lifelong use require battery replacement multiple times. A deep brain stimulator for Parkinson's disease treatment is implanted inside the human brain for neural signal sensing and stimulation [70]. Replacing the battery of the stimulator requires surgery on the patient that causes unnecessary inconvenience as well as safety issues. This problem can be overcome by using wireless charging based on inductive coupling. A deep brain stimulator with the wireless charging module implanted inside the human body [71] is shown in figure 7.15(a). However, in order to satisfy biocompatibility and safety issues for the wireless charging module in an implantable biomedical device, a titanium case is used sometimes for the protection of non-biocompatible electronic components in the module [72].

Wireless power transmission for biomedical applications is preferred at a carrier frequency close to 13.5 MHz because of less energy absorption and very small

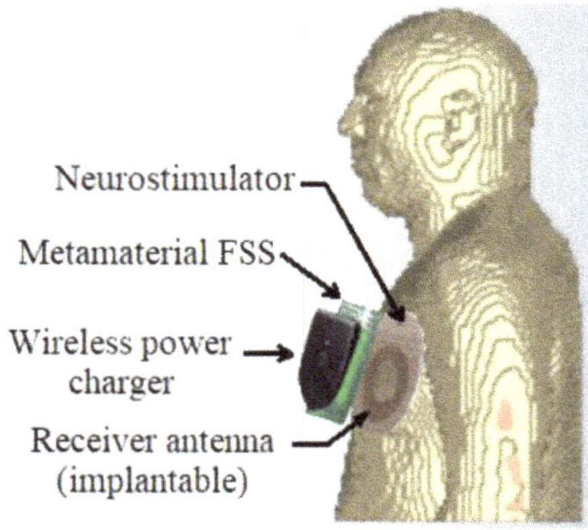

Figure 7.15. Metamaterial based FSS in deep brain stimulation system for Parkinson's disease.

amounts of tissue damage due to body penetration [73, 74]. However, the coupling efficiency of an implantable wireless power transmission (WPT) system in the near-field gets reduced with distance. Moreover, the eddy current loss due to metal packaging and misalignment between the antennas for transmission and reception in the body result in a poor efficiency for the wireless charging system. MTM-based periodic structures such as FSSs can be used to overcome these challenges [75, 76] as MTMs can enhance the evanescent waves with *E* and *H* fields in a specific orientation.

Pokharel et al. proposed [77] a WPT system using a metamaterial inspired geometry with near-zero permeability property, which overcame the performance degradation of WPT system in the time of biomedical implants. This metamaterial inspired geometry was stacked split ring resonator metamaterial fed by a driving inductive loop and acted as a WPT transmitter for an in-tissue implanted WPT receiver. The detailed figure of conventional WPT system and proposed stacked meta- surface WTP system inside the tissue are shown in figure 7.16(a). Layout of the convention design of the WTP transmitter and receiver unit cell as well as the proposed design is clearly shown with respective dimension in figure 7.16(b–d).
Chen *et al* proposed [77] a split ring resonator-shaped MTM array resonating at 13.56 MHz with a negative refractive index for enhancement of signal strength in a wireless power transfer (WPT) system used for implantable biomedical devices. The MTM unit cell consists of a double-sided metallic square spiral structure based on the FR4 substrate as shown in figure 7.16(a). The power transmission efficiency of the MTM-based WPT is calculated using *Z*-parameters obtained from measured *S*-parameters of transmitting or receiving coil antennas. For the measurement the MTM is placed along with the transmitting antenna as shown in figure 7.16(b). A 5%–7% increase in the WPT efficiency was observed with the use of MTM. The measurements were performed by keeping the misalignment (figure 7.16(b)) as well as misorientation (figure 7.16(c)) of the antennas to mimic the actual environment inside the human body while using the WPT system.

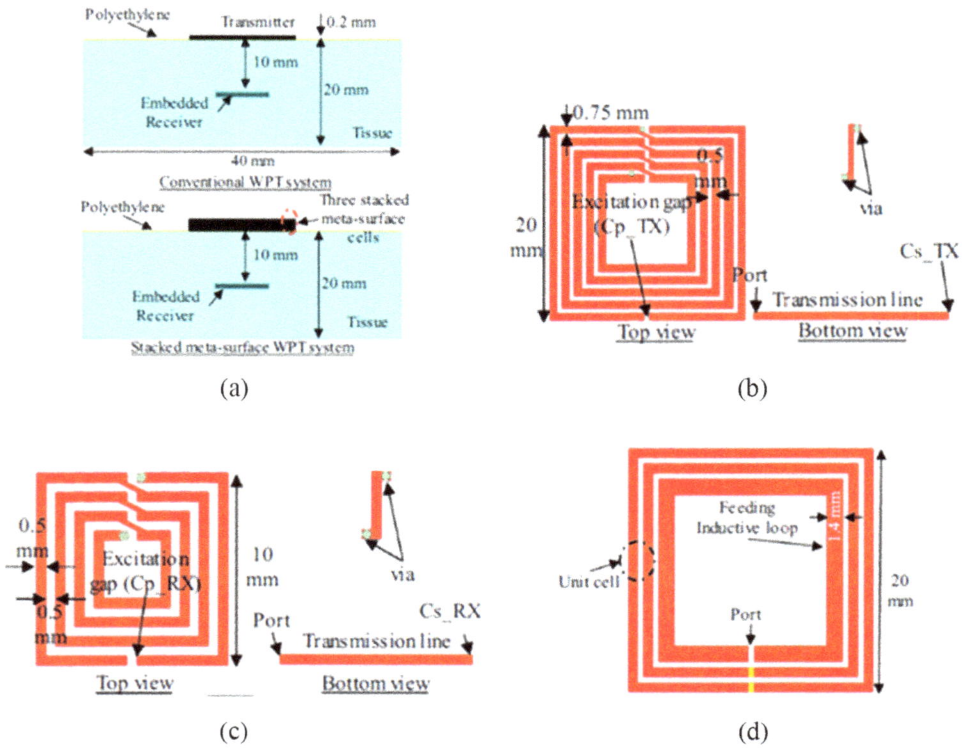

Figure 7.16. Conventional and proposed stacked metamaterial WPT systems in tissue. (a) Side view. (b) Layout of conventional TX . (c) Layout of the RX in both cases. (d) Layout of meta-TX. Reproduced from [77]. CC BY 4.0.

7.3.3 Integration with wearable antennas

Antennas as an integral part of the body area network (BAN) are used in all types of biomedical devices along with sensors and power sources. Wearable antennas are primarily used for on-body-centric communication where the antennas are mounted on the clothes worn on different parts of the human body, as was discussed earlier. However, wearable antennas generally exhibit poor gain that sometimes can also be negative [28, 78]. Apart from this, in the case of monopole antenna or bidirectional radiators, the radiation can affect tissues beneath the skin and continuous radiation may be harmful for the patient [31]. In such cases periodic structures such as EBG, FSS, and AMC, due to their band gap response and reflective nature, can be integrated with the antennas as substrate in order to reduce the back radiation [35]. FSS structures with bandpass response might also be used as a superstrate above the antenna to increase its gain and directivity in the broadside direction only [79–81]. However, due to array structure they increase the antenna dimension. Moreover, the spacing between antenna and FSS structures needs to be maintained at the conventional half-wavelength or quarter-wavelength value that in turn increases the profile further.

Metamaterials with negative permittivity and/or negative permeability, possessing unusual properties, are used to resolve the above-mentioned challenges by designing FSSs based on metamaterials. MTM-based periodic structures may exhibit antenna miniaturization as well as increase antenna gain or directivity. Mandal *et al* proposed [27] a miniaturized circularly polarized antenna with a metamaterial-inspired FSS-based superstrate in the shape of a button similar to those used in wearable clothes, as shown in figure 7.17(a). Transparent acrylic material is chosen as the substrate for both the antenna and FSS operating at an ISM frequency of 5.25 GHz, as shown in figure 7.17(b). The triangular antenna patch with the slits exhibits circular polarization (CP) with axial ratio <3 dB in the 5.02–5.39 GHz band. A metamaterial-inspired split ring resonator (square) with the dimension ($\lambda_g/8$) much less compared to conventional $\lambda/2$ or $\lambda/4$ value is used in the FSS superstrate. The antenna to MTM-based FSS spacing was kept very small ($0.03\lambda_0$) leading to low profile of the antenna. Moreover, the antenna with FSS covers an area of only 15 mm^2 close to the dimension of a typical button used in clothes, leading to a miniaturization of 34%. It was observed that the MTM-inspired FSS enhances antenna gain by 3 dB without affecting the CP. The SAR distribution as shown in figure 7.17(c) was observed to be 0.7 W kg^{-1} much lower than the threshold limit of 1.6 W kg^{-1}.

Metamaterial-based FSSs can also be used with the wearable antennas directly mounted above the skin of the human body without any spacing. Das *et al* proposed a wearable patch antenna with grounded metamaterial for biomedical application at 2.4 GHz [82]. The omega-shaped MTM unit cell based on 1 mm thick felt fabric textile with $\varepsilon_r = 1.63$ and loss tangent 0.044 is shown in figure 7.18(a). The shape is inspired from split rings and the MTM exhibits ε near zero (ENZ) and μ very large (MVL) property. It is shown by the authors that the power density of the antenna in the broadside direction P(0) is proportional to the ratio of relative permeability and permittivity of the material as given by equation (7.1) [82].

$$P(0) = \frac{K_0^2 \eta_0}{8\pi^2} \eta_r^2 \text{ where } \eta_r = \sqrt{\frac{\mu_r}{\varepsilon_r}} \tag{7.1}$$

The MTM was observed to exhibit very high permeability close to 80 and near zero permittivity close to 0.5. The authors analyzed the antenna with a 3 × 3 MTM

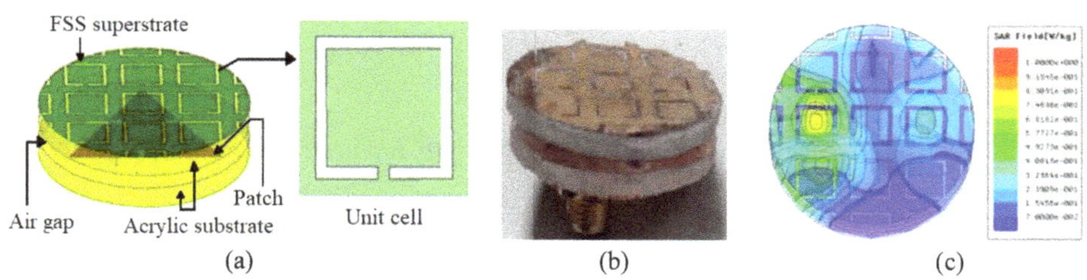

Figure 7.17. (a) Schematic of the button antenna with FSS superstrate, (b) fabricated button antenna with MTM-based FSS superstrate, and (c) SAR distribution for the antenna with FSS superstrate. [27] John Wiley & Sons. [© 2015 Wiley Periodicals, Inc.]

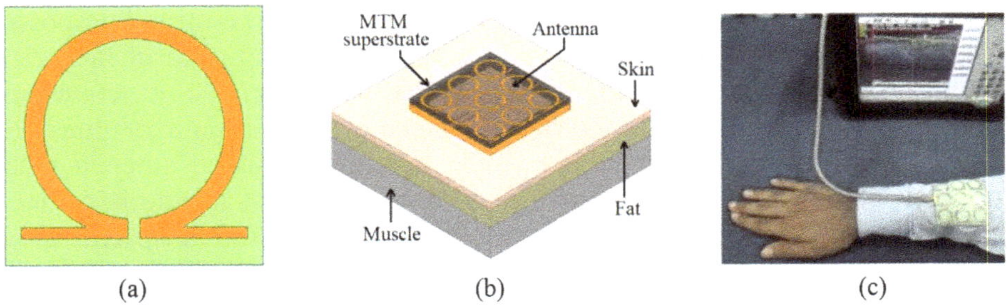

Figure 7.18. (a) Schematic of the metamaterial unit cell, (b) wearable patch antenna with MTM superstrate mounted on body phantom, and (c) fabricated MTM-loaded antenna on human arm. [82] John Wiley & Sons. [© 2020 The Institution of Engineering and Technology]

Table 7.2. Properties of the body phantom model. [82] John Wiley & Sons. [© 2020 The Institution of Engineering and Technology]

Body phantom	ε_r	σ (S m^{-1})	Thickness (mm)
Skin	38.1	1.43	2
Fat	5.29	0.1	10
Muscle	52.5	1.7	30

array on human body phantom model (figure 7.18(b)) with the properties of various layers beneath the skin as included in table 7.2.

It was observed that with the use of the MTM the antenna gain increases up to 8 dBi. The fabricated antenna was investigated with polyester, poly cot and wash cotton-based textile materials over human tissue as shown in figure 7.18(c) and a small shift in the resonance was observed. The antenna gain enhancement was 2.6 dB for the human body. The SAR value for the antenna-MTM is 0.405 W kg^{-1}, much below the threshold of 1.6 W kg^{-1} for 1 g tissue. Thus, it is evident that the use of metamaterial-inspired periodic structures in the antennas intended for biomedical applications leads to adequate enhancement in antenna gain as well as reduction of the back radiation as well as SAR.

References

[1] Enderle J and Bronzino J (ed) 2012 *Introduction to Biomedical Engineering* (New York: Academic)

[2] Lin J C 2004 Biomedical applications of electromagnetic engineering *Handbook of Engineering Electromagnetics* (Boca Raton, FL: CRC Press) 617–42

[3] Liu C, Guo Y X and Xiao S 2016 A review of implantable antennas for wireless biomedical devices *Forum for Electromagnetic Research Methods and Application Technologies (FERMAT)* **14** 1–11

[4] Abbasi Q H, Yang K, Chopra N, Jornet J M, Abuali N A, Qaraqe K A and Alomainy A 2016 Nano-communication for biomedical applications: a review on the state-of-the-art from physical layers to novel networking concepts *IEEE Access* **4** 3920–35

[5] Kiourti A, Psathas K A and Nikita K S 2014 Implantable and ingestible medical devices with wireless telemetry functionalities: a review of current status and challenges *Bioelectromagnetics* **35** 1–5

[6] Kiourti A and Nikita K S 2012 Accelerated design of optimized implantable antennas for medical telemetry *IEEE Antennas Wirel. Propag. Lett.* **11** 1655–8

[7] Kiourti A and Nikita K S 2013 Design of implantable antennas for medical telemetry: dependence upon operation frequency, tissue anatomy, and implantation site *Int. J. Monit. Surveillance Technol. Res.* **1** 16–33

[8] Kumar V, Ahmad M and Sharma A K 2010 Harmful effects of mobile phone waves on blood tissues of the human body *East. J. Med.* **15** 80

[9] Azzam E I, Colangelo N W, Domogauer J D, Sharma N and De Toledo S M 2016 Is ionizing radiation harmful at any exposure? An echo that continues to vibrate *Health Phys.* **110** 249–51

[10] Chatterjee A and Parui S K 2016 Performance enhancement of a dual-band monopole antenna by using a frequency-selective surface-based corner reflector *IEEE Trans. Antennas Propag.* **64** 2165–71

[11] Kundu S and Chatterjee A 2021 Sharp triple-notched ultra wideband Antenna with gain augmentation using FSS for ground penetrating radar *Wirel. Pers. Commun.* **117** 1399–418

[12] Devarapalli A B and Moyra T 2023 Design of a metamaterial loaded W-shaped patch antenna with FSS for improved bandwidth and gain *Silicon* **15** 2011–24

[13] Leelatien P, Ito K, Saito K, Sharma M and Alomainy A 2018 Channel characteristics and wireless telemetry performance of transplanted organ monitoring system using ultrawide-band communication *IEEE J. Electromagn. RF Microw. Med. Biol.* **2** 94–101

[14] Valdastri P, Menciassi A, Arena A, Caccamo C and Dario P 2004 An implantable telemetry platform system for in vivo monitoring of physiological parameters *IEEE Trans. Inf. Technol. Biomed.* **8** 271–8

[15] Jovanov E, Milenkovic A, Otto C, De Groen P, Johnson B, Warren S and Taibi G 2006 A WBAN system for ambulatory monitoring of physical activity and health status: applications and challenges *2005 IEEE Engineering in Medicine and Biology 27th Annual Conf.* (Piscataway, NJ: IEEE) 3810–3

[16] Zerrouki H and Azzaz-Rahmani S 2021 Design of dual-band (MICS and ISM) implantable antenna for wireless medical telemetry applications *J. Electr. Syst.* **17** 350–62

[17] Ung J and Karacolak T 2012 A wideband implantable antenna for continuous health monitoring in the MedRadio and ISM bands *IEEE Antennas Wirel. Propag. Lett.* **11** 1642–5

[18] González-Molero I *et al* 2012 Use of telemedicine in subjects with type 1 diabetes equipped with an insulin pump and real-time continuous glucose monitoring *J. Telemed. Telecare* **18** 328–32

[19] Vithayathil M, Prabhu B, Ningthoujam R and Singh L S 2019 Designing and modelling of a low-cost wireless telemetry system for deep brain stimulation studies *Indian J. Sci. Technol.* **12** 1–13

[20] Pellegrini A *et al* 2013 Antennas and propagation for body-centric wireless communications at millimeter-wave frequencies: a review [wireless corner] *IEEE Antennas Propag. Mag.* **55** 262–87

[21] Li H B, Takizawa K I, Zhen B and Kohno R 2007 Body area network and its standardization at IEEE 802.15. MBAN *2007 16th IST Mobile and Wireless Communications Summit* (Piscataway, NJ: IEEE) 1–5

[22] Abbas S M, Esselle K P, Matekovits L, Rizwan M and Ukkonen L 2016 On-body antennas: Design considerations and challenges *2016 URSI Int. Symp. on Electromagnetic Theory (EMTS)* (Piscataway, NJ: IEEE) 109–10

[23] Singh S and Prasad D 2022 Wireless body area network (WBAN): a review of schemes and protocols *Mater. Today Proc.* **49** 3488–96

[24] Mandal B and Parui S K 2015 A miniaturized wearable button antenna for Wi-Fi and Wi-Max application using transparent acrylic sheet as substrate *Microw. Opt. Technol. Lett.* **57** 45–9

[25] Mandal B and Parui S K 2015 Wearable tri-band SIW based antenna on leather substrate *Electron. Lett.* **51** 1563–4

[26] Le T T and Yun T Y 2020 Miniaturization of a dual-band wearable antenna for WBAN applications *IEEE Antennas Wirel. Propag. Lett.* **19** 1452–6

[27] Mandal B, Chatterjee A and Parui S K 2015 Acrylic substrate based low profile wearable button antenna with FSS layer for WLAN and Wi-Fi applications *Microw. Opt. Technol. Lett.* **57** 1033–8

[28] El Gharbi M, Martinez-Estrada M, Fernández-García R, Ahyoud S and Gil I 2021 A novel ultra-wide band wearable antenna under different bending conditions for electronic-textile applications *J. Tex. Inst* **112** 437–43

[29] Smith D R, Padilla W J, Vier D C, Nemat-Nasser S C and Schultz S 2000 Composite medium with simultaneously negative permeability and permittivity *Phys. Rev. Lett.* **84** 4184

[30] Vendik I B and Vendik O G 2013 Metamaterials and their application in microwaves: a review *Tech. Phys.* **58** 1–24

[31] Salonen P, Rahmat-Samii Y and Kivikoski M 2004 Wearable antennas in the vicinity of human body *IEEE Antennas and Propagation Society Symp.* **vol 1** (Piscataway, NJ: IEEE) 467–70

[32] Tuovinen T, Berg M and Salonen E T 2014 Antenna close to tissue: avoiding radiation pattern minima with an anisotropic substrate *IEEE Antennas Wirel. Propag. Lett.* **13** 1680–3

[33] Ali U, Ullah S, Kamal B, Matekovits L and Altaf A 2023 Design, analysis and applications of wearable antennas: a review *IEEE Access* **11** 14458–86

[34] Kapoor A, Mishra R and Kumar P 2022 Frequency selective surfaces as spatial filters: fundamentals, analysis and applications *Alex. Eng. J.* **61** 4263–93

[35] Mandal B, Chatterjee A, Rangaiah P, Perez M D and Augustine R 2020 A low profile button antenna with back radiation reduced by FSS *2020 14th European Conf. on Antennas and Propagation (EuCAP)* (Piscataway, NJ: IEEE) pp 1–5

[36] Bayatpur F and Sarabandi K 2008 Multipole spatial filters using metamaterial-based miniaturized-element frequency-selective surfaces *IEEE Trans. Microw. Theory Tech.* **56** 2742–7

[37] Kai Z, Soh Ping J and Sen Y 2020 Meta-wearable antennas—a review of metamaterial based antennas in wireless body area networks *Materials* **14** 149

[38] Roy S and Chakraborty U 2019 Metamaterial based dual wideband wearable antenna for wireless applications *Wirel. Pers. Commun.* **106** 1117–33

[39] AlSabbagh H M, Elwi T A, Al-Naiemy Y and Al-Rizzo H M 2020 A compact triple-band metamaterial-inspired antenna for wearable applications *Microw. Opt. Technol. Lett.* **62** 763–77

[40] Al-Adhami A and Ercelebi E 2021 A flexible metamaterial based printed antenna for wearable biomedical applications *Sensors* **21** 7960
[41] Saeidi T, Ismail I, Mahmood S N, Alani S, Ali S M and Alhawari A R 2020 Metamatrial-based antipodal vivaldi wearable UWB antenna for IoT and 5G applications *2020 IEEE Int. Conf. on Dependable, Autonomic and Secure Computing, Int. Conf. on Pervasive Intelligence and Computing, Int. Conf. on Cloud and Big Data Computing, Int. Conf. on Cyber Science and Technology Congress (DASC/PiCom/CBDCom/CyberSciTech)* (Piscataway, NJ: IEEE) 14–20
[42] Ali Khan M U, Raad R, Tubbal F, Theoharis P I, Liu S and Foroughi J 2021 Bending analysis of polymer-based flexible antennas for wearable, general IoT applications: a review *Polymers* **13** 357
[43] Dhanabalan S S, Sitharthan R, Madurakavi K, Thirumurugan A, Rajesh M, Avaninathan S R and Carrasco M F 2022 Flexible compact system for wearable health monitoring applications *Comput. Electr. Eng.* **102** 108130
[44] Noor S and Ramli N 2020 A review of the wearable textile-based antenna using different textile materials for wireless applications *Open J. Sci. Technol.* **3** 237–44
[45] Soh P J, Vandenbosch G A, Wee F H, Van Den Bosch A, Martinez-Vazquez M and Schreurs D P 2013 Specific absorption rate (SAR) evaluation of biomedical telemetry textile antennas In *2013 IEEE MTT-S Int. Microwave Symp. Digest (MTT)* (Piscataway, NJ: IEEE) pp 1–3
[46] Kamal B, Ali U, Chen J and Ullah S 2022 Applications of metamaterials and metasurfaces *Metamaterials-History, Current State, Applications, and Perspectives* (London: IntechOpen)
[47] Gil I, Seager R and Fernández-García R 2018Nov Embroidered metamaterial antenna for optimized performance on wearable applications *Phys. Status Solidi (a)* **215** 1800377
[48] Alhawari A R, Almawgani A H, Hindi A T, Alghamdi H and Saeidi T 2021 Metamaterial-based wearable flexible elliptical UWB antenna for WBAN and breast imaging applications *AIP Adv.* **11** 015128
[49] Ramasamy K, Sapna B A and Jayasheela M 2023 A novel wearable monopole antenna with controlled SAR using metamaterial *Int. J. Microw. Wireless Technol.* **15** 1524–36
[50] Koydemir H C and Ozcan A 2018 Wearable and implantable sensors for biomedical applications *Annu. Rev.Anal. Chem.* **11** 127–46
[51] Chow E Y, Morris M M and Irazoqui P P 2013 Implantable RF medical devices: the benefits of high-speed communication and much greater communication distances in biomedical applications *IEEE Microwave Mag.* **14** 64–73
[52] Nag S and Thakor N V 2016 Implantable neurotechnologies: electrical stimulation and applications *Med. Biol. Eng. Comput.* **54** 63–76
[53] Wang W, Xuan X W, Zhao W Y and Nie H K 2021 An implantable antenna sensor for medical applications *IEEE Sens. J.* **21** 14035–42
[54] Bhattacharjee S, Maity S, Bhadra Chaudhuri S R and Mitra M 2018 Metamaterial-inspired wideband biocompatible antenna for implantable applications *IET Microw. Antennas Propag.* **12** 1799–805
[55] Zada M, Shah I A and Yoo H 2019 Metamaterial-loaded compact high-gain dual-band circularly polarized implantable antenna system for multiple biomedical applications *IEEE Trans. Antennas Propag.* **68** 1140–4
[56] Saidi A, Nouri K, Bouazza B S, Becharef K, Cherifi A and Abes T 2022 E-shape metamaterials embedded implantable antenna for ISM-band biomedical applications *Res. Biomed. Eng* **38** 351–68

[57] Keltouma N, Amaria S, Khadidja B, Mokhtaria C, Salima S, Kada B and Turkiya A 2023 Dual-band half-circular ring implantable antenna with metamaterial SRR For biomedical applications *Telecommun. Radio Eng.* **82** 69–86
[58] Khan A N, Cha Y-O, Giddens H and Hao Y 2022 Recent advances in organ specific wireless bioelectronic devices: perspective on biotelemetry and power transfer using antenna systems *Engineering* **11** 27–41
[59] Xu Y, Hu X, Kundu S, Nag A, Afsarimanesh N, Sapra S, Mukhopadhyay S C and Han T 2019 Silicon-based sensors for biomedical applications: a review *Sensors* **19** 2908
[60] Maduraiveeran G, Sasidharan M and Ganesan V 2018 Electrochemical sensor and biosensor platforms based on advanced nanomaterials for biological and biomedical applications *Biosens. Bioelectron.* **103** 113–29
[61] Emami-Nejad H and Mir A 2017 Design and simulation of a flexible and ultra-sensitive biosensor based on frequency selective surface in the microwave range *Opt. Quantum Electron.* **49** 1–5
[62] Chandra P, Tan Y N and Singh S P (ed) ed *Next Generation Point-Of-Care Biomedical Sensors Technologies for Cancer Diagnosis* (Singapore: Springer) 2017
[63] Henry N L and Hayes D F 2012 Cancer biomarkers *Mol. Oncol* **6** 140–6
[64] Banerjee S, Dutta P, Jha A V, Appasani B and Khan M S 2022 A biomedical sensor for detection of cancer cells based on terahertz metamaterial absorber *IEEE Sens. Lett.* **6** 1–4
[65] Wang L 2018 Microwave sensors for breast cancer detection *Sensors* **18** 655
[66] Geng Z, Zhang X, Fan Z, Lv X and Chen H 2017 A route to terahertz metamaterial biosensor integrated with microfluidics for liver cancer biomarker testing in early stage *Sci. Rep.* **7** 16378
[67] Senapati S, Mahanta A K, Kumar S and Maiti P 2018 Controlled drug delivery vehicles for cancer treatment and their performance *Signal Transduct. Target. Ther* **3** 7
[68] Zisser H, Robinson L, Bevier W, Dassau E, Ellingsen C, Doyle III F J and Jovanovic L 2008 Bolus calculator: a review of four 'smart' insulin pumps *Diabetes Technol. Ther* **10** 441–4
[69] Goodman A M, Neumann O, Nørregaard K, Henderson L, Choi M R, Clare S E and Halas N J 2017 Near-infrared remotely triggered drug-release strategies for cancer treatment *Proc. Natl Acad. Sci.* **114** 12419–24
[70] Okun M S 2012 Deep-brain stimulation for Parkinson's disease *New Engl. J. Med.* **367** 1529–38
[71] Niroomand M and Lotfian M 2015 A wireless based power transmission to supply deep brain stimulators *J. Biomed. Res.* **26** 497–504
[72] Nathan M 2010 Microbattery technologies for miniaturized implantable medical devices *Curr. Pharm. Biotechnol.* **11** 404–10
[73] Ibrahim A, Meng M and Kiani M 2018 A comprehensive comparative study on inductive and ultrasonic wireless power transmission to biomedical implants *IEEE Sens. J.* **18** 3813–26
[74] Machnoor M, Rodríguez E S, Kosta P, Stang J and Lazzi G 2018 Analysis and design of a 3-coil wireless power transmission system for biomedical applications *IEEE Trans. Antennas Propag.* **67** 5012–24
[75] Shuai K, Yang J, Liu C, Yang X and Liu X 2021 Beam-steering transmitarray with active frequency selective surface for wireless power transmission *Int. J. RF Microw. Comput.-Aided Eng* **31** e22907

[76] Corrêa D C, Resende U C and Bicalho F S 2019May22 Experiments with a compact wireless power transfer system using strongly coupled magnetic resonance and metamaterials *IEEE Trans. Magn.* **55** 1–4
[77] Pokharel R K, Adel B, Shimaa A, Kuniaki Y and Costas S 2021 Wireless power transfer system rigid to tissue characteristics using metamaterial inspired geometry for biomedical implant applications *Sci. Rep.* **11** 5868
[78] Khajeh-Khalili F, Shahriari A and Haghshenas F 2021 A simple method to simultaneously increase the gain and bandwidth of wearable antennas for application in medical/communications systems *Int. J. Microw. Wireless Technolog.* **13** 374–80
[79] Chatterjee A and Parui S K 2015 Gain enhancement of a wide slot antenna using a second-order bandpass frequency selective surface *Radioengineering* **24** 455–61
[80] Rana B, Chatterjee A and Parui S K 2016 Gain enhancement of a dual-polarized dielectric resonator antenna using polarization independent FSS *Microwave Opt. Technol. Lett.* **58** 1415–20
[81] Ballav S, Chatterjee A and Parui S K 2021 Gain augmentation of a dual-band dielectric resonator antenna with frequency selective surface superstrate *Int. J. RF Microw. Comput.-Aided Eng* **31** e22575
[82] Das G K, Basu S, Mandal B, Mitra D, Augustine R and Mitra M 2020 Gain-enhancement technique for wearable patch antenna using grounded metamaterial *IET Microw. Antennas Propag.* **14** 2045–52

Chapter 8

3D metamaterial: recent trends and applications

8.1 Introduction

The emergence of metamaterials is thought to have caused one of the greatest revolutions in electromagnetics this century [1]. Metamaterials are artificial composite constructions created by humans that enable the manipulation of electromagnetic and acoustic wave propagation in various mediums. These artificial structures go beyond the properties of their constitutive materials; namely, the geometrical disposal of their elements (frequently named 'meta-atoms') largely determines their mechanical or electrical response, like crystals and protein chains. The most outstanding point about metamaterials is that they can be engineered to exhibit unusual properties rarely found in nature [2, 3], i.e., artificial magnetism (magnetism without inherent magnetic materials), negative-refractive indexes from positive-index materials, invisibility cloaking ('invisible' materials that do not interact with light), non-reciprocal phenomena, and chiral responses, such as the one schematized in figure 8.1.

The basic difference between metamaterials and metasurfaces is their structure. In essence, metamaterials are artificial materials made of metals and/or dielectrics that are three-dimensional (3D) and frequently periodic. Because of their high interaction with electric and/or magnetic fields which is usually given by resonant effects regulated by the geometry of the unit cells they can manipulate waves with outstanding effectiveness. Numerous applications, including improving antenna performance [5, 6], perfect absorbers [7–9], super lenses [10, 11], cloaking [12–14], scattering reduction [15], and energy harvesting [16–18], are made possible by the characteristics of 3D metamaterials. However, many applications of metamaterials are impeded by high losses and difficulties in 3D fabrication, especially on micro- and nano-scales. Metasurfaces are flat or two-dimensional (2D) metamaterials that usually have subwavelength thickness [19–24]. They are broadly investigated and implemented in electromagnetics applications due to their light weight and ease of

doi:10.1088/978-0-7503-5422-6ch8

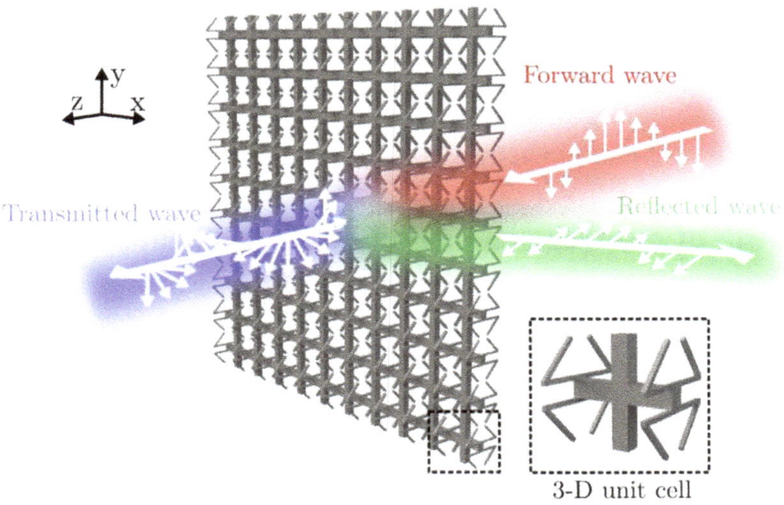

Figure 8.1. A conceptual illustration of a 3D metamaterial controlling the flow of light. Adapted from [4] CC BY 4.0.

fabrication. Metasurfaces are unusual in their ability to block, absorb, concentrate, disperse, or guide waves from microwave through optical frequencies, both at grazing incidence on the surface and at normal and oblique incidence in space.

By creating metasurface unit cells with the necessary impedance to regulate the phase or group velocity, EM waves can be controlled simultaneously [25]. They are employed in scattering control applications and have patterns that can split or steer waves in specific directions. It is possible to achieve a variety of effective surface refractive indices and pattern the surface to serve a variety of purposes by adjusting the sizes and forms of the metasurface unit cells. They can be used to create 2D optical and microwave lenses, such as Luneburg and fish-eye lenses, which are employed in planar microwave sources and surface waveguides for antenna systems. These materials have potential uses in optical communication and imaging because they have the ability to refract light in the opposite way to normal materials. By analogy with EM metamaterials that can control EM wave propagation, researchers proposed acoustic metamaterials that could be used to manipulate acoustic waves [26, 27]. Various acoustic performances can be achieved based on acoustic metamaterials, including acoustic cloaking/attenuation [28], and acoustic lenses with directional acoustic wave transmission functions. Another crucial topic in metamaterials is achieving unusual mechanical properties [29, 30]. This involves using periodic structures to obtain desired stiffness, flexibility, sensing, and shape morphing capabilities [31, 32]. For example, negative Poisson's ratio [33, 34], negative thermal expansion [35], bi-stability [36, 37]. In addition to their mechanical and optical properties, 3D printed metamaterials have also been explored for their thermal properties [38], which can be applied in thermal cloaking, thermal concentration, and thermal rotation. The development of 3D printed metamaterials has opened a wide range of possibilities for creating materials with novel and highly customizable properties. While there is still much to explore in this field, the potential applications of 3D printed metamaterials are vast, ranging from biomedical devices and sensors to energy harvesting and storage.

In this chapter, we provide an overview of recent developments in the field of metasurfaces from passive to active, from microwave through visible frequencies.

8.2 Growth of research and market of 3D metamaterials

In 2019, Vicari *et al* [39] proposed the forecasting of metamaterials' market. Based on a few factors, such as cost, maturity, and performance, the model examined prospective addressable markets. They clearly showed that throughout the years, the communication and sensing sector will be heavily dependent on metamaterial-based products.

When compared to one- and two-dimensional metasurfaces, three-dimensional metamaterials have advantages [40]. These benefits are generally ascribed to the fact that a 3D structure makes it possible to utilize the degrees of freedom in both the longitudinal and transverse directions during the design process. Put differently, there are more geometrical parameters to adjust the device's behavior. Thus, the advanced functionalities in 3D metamaterials are difficult to achieve with 1D and 2D configurations. This can be advantageous for smart radio settings and future communication systems [41]. When two orthogonal polarization states are independently controlled, a structure can work as both a wave absorber and a beam scanner (splitter/steerer) at the same time, demonstrating sophisticated capability [42]. Additionally, the utilization of totally metallic implementations rather than dielectric materials might enhance the efficiency in 3D structures. [43]. Planar structures made using conventional printed circuit board (PCB) processes, in which the metasurface is bonded to a dielectric substrate, are unable to achieve this. Additionally, if fully metallic designs are taken into consideration, scaling 3D structures to different frequency ranges can be done with ease.

Naturally, 2D or 1D metasurfaces are more cost effective than 3D metamaterials. The bulkier nature of 3D metamaterials compared to planar structures may be prohibitive for some applications. Additionally, they are far more difficult to make and need more processing power, which has hitherto prevented their widespread use. This trend has changed in recent years, thanks to the massive technological progress in computer electronics (increased computational power) and the development of 3D printing techniques. Some different 3D metamaterial designs for different fields (mechanics, acoustics, and electromagnetics) are illustrated in figure 8.2.

8.3 Some computational methods for 3D structures solution in recent years

There are some computational methods in recent activities on 3D metamaterial processes. Every approach has advantages and disadvantages that determine in which areas it works best in and which ones it doesn't [50].

Finite-difference method (FDM) approach: This is the most basic full-wave method for solving all of Maxwell's equations [51, 52]. Finite-difference time-

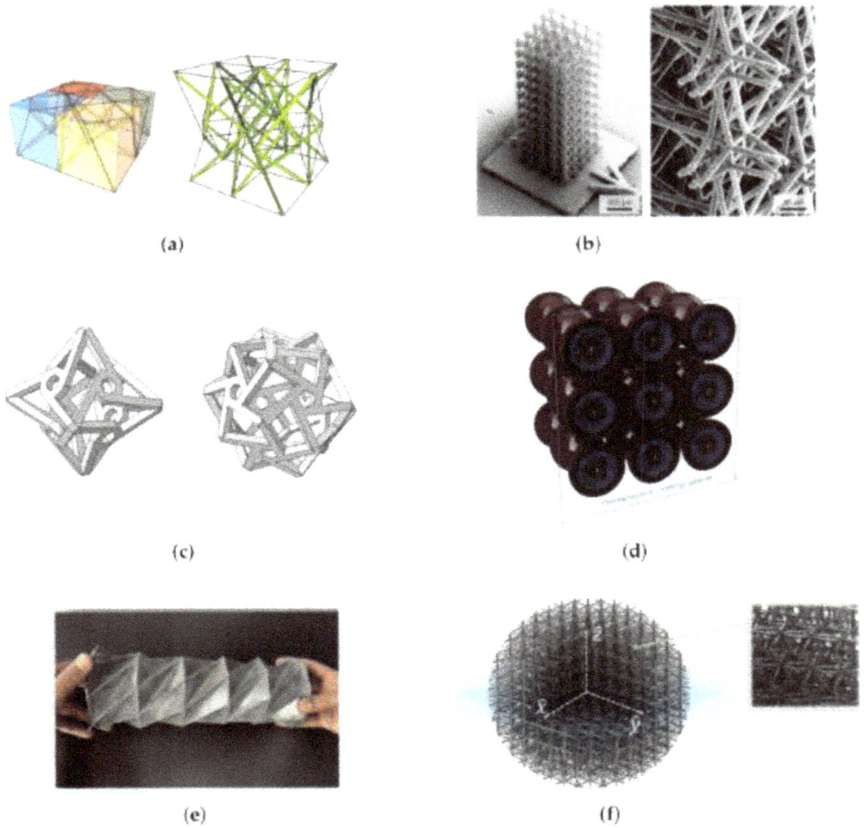

Figure 8.2. Examples of 3D metamaterials. Mechanical metamaterials. (a) Four- and eight-module supercells. Reproduced from [44]. CC BY 4.0. (b) Chiral cubic polymer metamaterial. Reproduced from [45]. CC BY 4.0. (c) Chiral pyramid and dodecahedron lattice. Reproduced from [46]. CC BY 4.0. (d) Acoustic metamaterial. Reproduced from [47]. CC BY 4.0. (e) Antenna. Reproduced from [48]. CC BY 4.0. (f) Electromagnetic metamaterial. Reproduced from [49]. CC BY 4.0.

domain (FDTD) approach: Yee in 1966 [53] found out the transformation method of the original Maxwell's equations into difference equations to solve a variety of 3D electromagnetic problems.

Finite-difference frequency-domain (FDFD) approach: FDTD is normally preferable to analyze transient and nonlinear phenomena while FDFD is a better option for electrically large structures. As an example, FDTD schemes are implemented in [54–56] for the analysis of 3D (nano) cavities and woodpile structures, respectively. FDFD techniques also have been implemented for modelling the scattering phenomena in isotropic [57] and anisotropic [58] 3D objects, and the dispersion properties of open nanophotonic resonators [59].

Finite element method (FEM) approach: Like FDTD and FDFD, the whole space in FEM is discretized into small pieces, in this case, called finite elements. In the literature, there are some examples of FEM formulations used in electromagnetic 3D design. FEM was created to build efficient cloaks for spheroidal metallo-dielectric structures in [60]. In [61] EM fields in 3D nanophotonic crystals and waveguides were computed using this method. The commercial EM simulators like COSMOL, CST, Ansys HFSS, etc., use FEM techniques for theoretical simulation process [62, 63].

Method of Moments (MoM) approach: The most famous integral equation, occasionally named the boundary element method in other fields of physics [64]. MoM has historically been used to investigate wire and planar antennas, single-layer reflect array/transmit array topologies, planar multilayered (2.5D) devices [65–67]. Generally, this method is not fully supportable for 3D geometry for substantial time complexity, though MoM formulations are developed in [68] (combined with a plane-wave expansion) for an analysis of 3D metallic wire media and in [69].

Besides the above mentioned method, some other methods are also used in this 3D structural field. The Nyström Method, an alternative to MoM, used to find out scattering phenomena in 3D objects and composite structures [70, 71]. Another method, Fast Multipole Method (FMM) is an integral-equation approach widely utilized for solving gravitational and electromagnetic issues [72]. In [73], the scattering of electromagnetic waves in 3D nanoparticle systems is studied using a hybrid FMM-MoM approach. The semi-infinite 3D plasmonic media presented in [74], the Wiener–Hopf method, which is a powerful technique for solving certain types of integral equations by making use of the Laplace transform (Z transform in discrete implementations), has been used to find insightful and efficient solutions. Ewald's method has been typically applied to compute the dispersion properties in 3D periodic arrays of electric and magnetic dipoles [75, 76] and in 3D skewed lattices [77].

Circuit modelling is another insightful tool which can make the solution of a electromagnetic surface a physical reality. By this modelling method, a unit structure of a 3D metamaterial can be expressed in a circuit of resistive, capacitive, and inductive lumped elements. In [78] the metamaterial has been solved by an equivalent circuit model (ECM). A heuristic circuit technique was used to investigate a periodic 3D metamaterial constructed by laterally perforated square waveguides that exploit phase resonances in [79, 80].

8.4 Some designs of 3D metamaterial-based structures

As mentioned above, in recent times complicated structures of 3D metamaterial have been developed by the researchers. The basic metamaterial prototype was a 3D periodic distribution of split-ring resonators that performed electrical behaviors not found in nature and were easily programmable. Many researchers focused on these unusual features, resulting in the creation of prospective applications such as those in the field of magnetic resonances [81]. Using homogeneous techniques, the model analysis of 3D split ring structure has been reported in reference [82]. In recent times, researchers have been using artificial photonic crystal formed by fully metallic lattices of parallel rods [83] or interconnected wires under exotic geometries [84, 85] for making the 3D metamaterial structure work in the THz frequency range.

In 2021, Saha *et al* proposed a 3D metamaterial-based FSS structure, which provided tunable frequency behavior depending on the internal air gap between two adjacent FSS layers [86]. In figure 8.4 the proposed 3D-FSS structure and its tunable frequency behaviors is shown clearly. Here, a three-layer metamaterial structure-based FSS has been arranged in cascaded format.

For the Ka-band, a completely metallic transmit array with good insertion losses and a wide bandwidth response was proposed in [89]. It's important to note that the inclusion of ground planes, which require the full reflection operation, makes the term 'fully dielectric' generally inaccurate. Hence, the most attractive designs are dielectric reflect arrays built on protrusion dielectric substrates. In some, examples of paradigmatic evidence are reported. The example with an elliptical shaped dielectric resonator (DR) is detailed in [90]. Cross-shaped protrusions and rectangular protrusions are also explored [91]. Regarding 3D lenses, it is important to draw attention to the prototypes of Luneburg and Fresnel lenses that were described in [92, 93]. The first one uses unconventional cubic cells to create the lens in the Ku/K-bands, while the second one optimizes the lens in the THz domain by combining air and dielectric slabs to adjust the phase of each cell. Both examples exhibit excellent gain performances and high efficiency.

8.5 Recent fabrication process of 3D metamaterial structures

In the case of recent 3D geometry of metamaterial structures, modern fabrication techniques and assembling are important for research. Even with the usage of metasurfaces to construct stacks, PCB manufacture was able to achieve an appropriate prototype in planar traditional fabrication, where metasurfaces have 2D geometry. These results, if applied, must be supplemented by more intricate assembly in the case of unit cell production with 3D geometry, as they are insufficient. This section carries out a review of several literary works that present a three-dimensional unit cell prototype. Generally, three types of fabrication process are followed to fabricate the modern 3D metamaterial geometries.

- **Conventional manufacturing techniques:** Any conventional manufacturing method, such as PCB manufacturing, CNC, or electrical discharge machining (EDM), can be used in this fabrication process. These methods involve the independent fabrication of the components that make up the prototype and their subsequent three-dimensional assembly, as depicted in figure 8.3. These types of fabrication process have been followed in [87, 94–96]. In these cases, different 2D PCB layers were stacked both horizontally and vertically to construct the sides of the 3D unit cells during manufacturing. These parts are assembled simply by utilizing several grooves that interlock with one another [95]. It is shown in figure 8.4(a). In some cases, conducting glue and soldering has been used to assemble these layers and avoid the electrical discontinuity of the connections [87]. A different approach to 3D assembly using PCB layers is shown in [97], where copper wires serve as the supporting and connecting elements between the two layers, thereby improving the electromagnetic responses. In figure 8.4(b), a fabricated 3D unit cell is shown where magnetic materials are inserted into the area of unit cells.
- **3D printing technique:** Another important fabricating technique is 3D printing. Within this category, prototypes that are 3D printed on plastic and then metallized can be distinguished from those that are 3D printed directly on

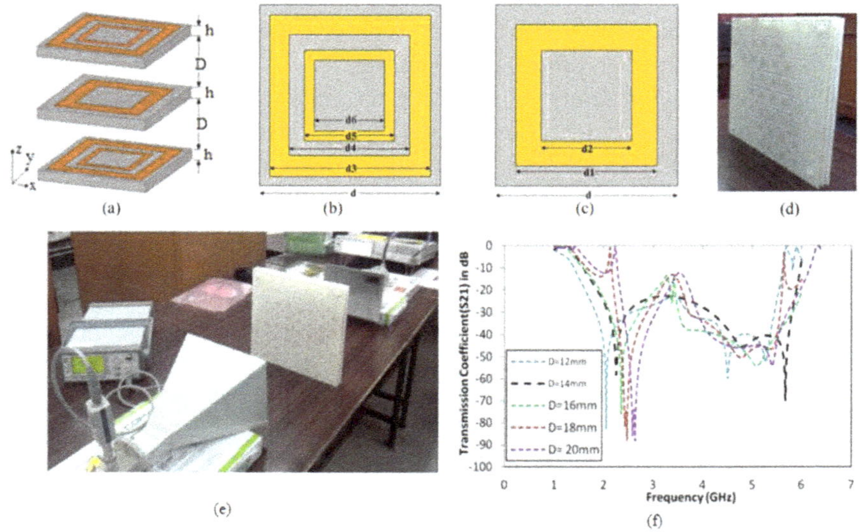

Figure 8.3. Examples of 3D metamaterial-based three-layer FSS structure. (a–c) Design of each layer. (d) Fabricated structure. (e) Measurement set up. (f) Reconfigurable frequency characteristics (all are adopted from reference. Reproduced from [89]. CC BY 4.0.

metal, which often don't need metallization afterwards. For general-purpose 3D printers, fused filament manufacturing, also known as fused deposition modeling (FDM), is the most popular method. The printed object is created through the fusion of a thermoplastic through the use of an extruder nozzle. The FDM 3D printers used in the prototypes described in [88] use thermoplastic filaments made of acrylonitrile butadiene styrene (ABS) and polylactic acid (PLA). The prototype from [88] is depicted in Figure 8.4(a), where a cross-section of the unit cell makes evident the manufacturing process. The material jetting (MJ) method of 3D printing, which is another technology covered in the aforementioned studies, works by first depositing material droplets in each layer, which are then cured by UV radiation before the next layer is produced. These types of 3D printing techniques have been reported in [98, 99] as shown in figure 8.5.

- **Other techniques:** Using water transfer printing (WPT) technology, a metasurface was effectively deposited into a three-dimensional metal mold. The metasurface was first created via inkjet printing on a flexible substrate, like polyethylene terephthalate (PET), before utilizing the WPT. Because of the geometry of the mold, WPT allows the unit cells to take on a 3D form. In [100], this technique has been followed by the researchers. Another method, followed by the researchers for prototyping of 3D metamaterial surface structures, was inject printing. In this method the flexible dielectric is molded into a 3D geometric unit cell structure. In [101], the cellulose paper has been formed into 3D-FSS by inject printing. Laser printing is another modern technique,which is used in reference [86]. In [102, 103], precise manufacturing of lithography techniques has been followed.

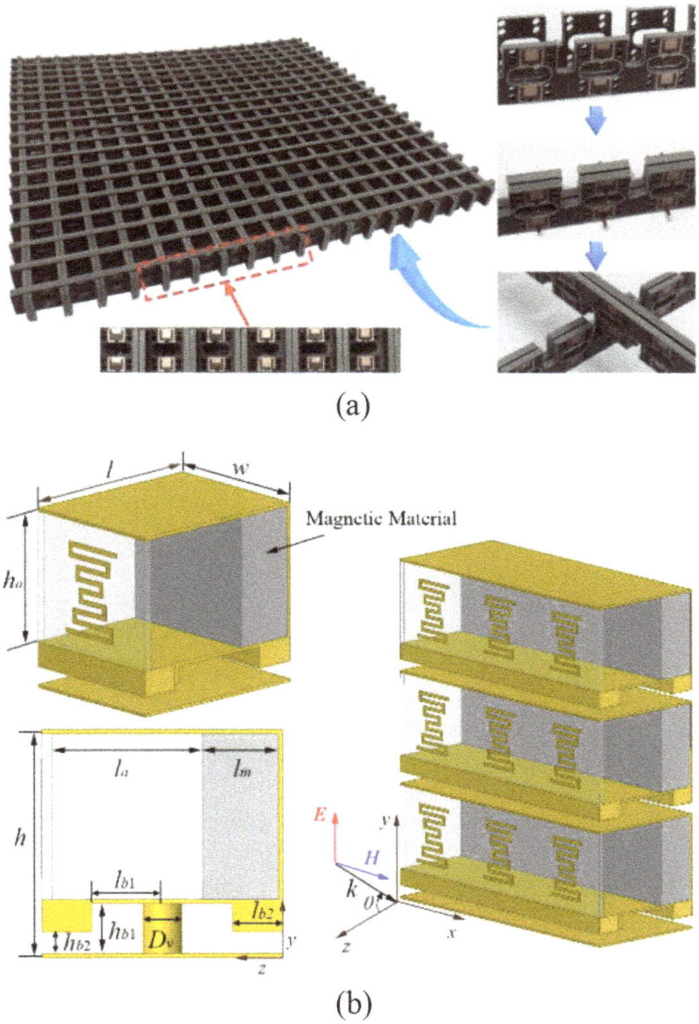

Figure 8.4. (a) Fabricated 3D metamaterial assembled with groove. Reproduced from [87]. CC BY 4.0. (b) 3D unit cell and periodic array structure inserted with magnetic materials. Reproduced from [88]. CC BY 4.0.

Over the chapter, the recent trends and modernization of 3D metamaterial-based designs have been explored. In the introductory part advantages of 3D metamaterials over 1D and 2D metamaterials have been discussed. In the next part, the recent and future market status of the metamaterial industry has been discussed. In the later sections, the mathematical simulation processes and recent fabrication processes of 3D geometrical structures have been explained, respectively. These modern types of metamaterials are very useful for reconfigurable intelligence surfaces (RISs), which can be useful for intelligence vehicles [104] and vortex-generation for RF imaging systems [105], etc.

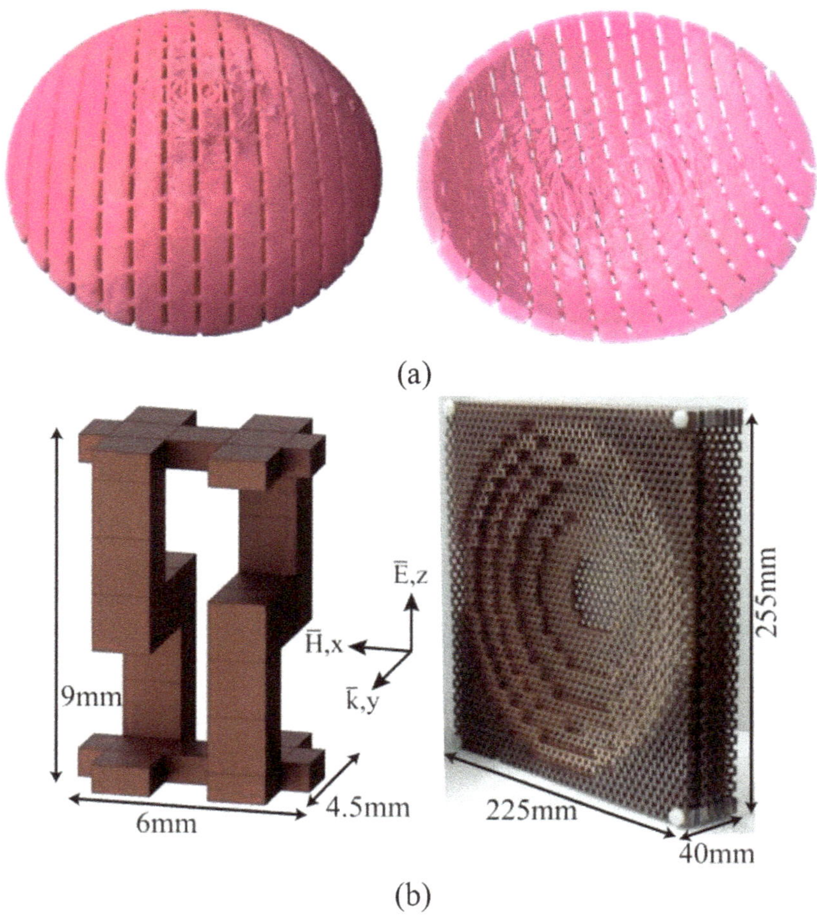

Figure 8.5. (a) Convex and concave sides of the 3D printed conformal FSS (adopted from reference [98]). (b) Solid model of the S-ring unit cell, materials and a photograph of the assembled planoconcave lens (adopted from reference [99]).

References

[1] Smith D R, Pendry J B and Wiltshire M C 2004 Metamaterials and negative refractive index *Science* **305** 788–92

[2] Engheta N and Ziolkowski R W (ed) 2006 *Metamaterials: Physics and Engineering Explorations* (New York: Wiley)

[3] Liu Y and Zhang X 2011 Metamaterials: a new frontier of science and technology *Chem. Soc. Rev.* **40** 2494–507

[4] Alex-Amor A, Palomares-Caballero Á and Molero C 2022 3-D metamaterials: trends on applied designs, computational methods, and fabrication techniques *Electronics* **11** 410

[5] Alici K B and Özbay E 2007 Radiation properties of a split ring resonator and monopole composite *Phys. Status solidi (B)* **244** 1192–6

[6] Enoch S, Tayeb G, Sabouroux P, Guérin N and Vincent P 2002 A metamaterial for directive emission *Phys. Rev. Lett.* **89** 213902
[7] Landy N I, Sajuyigbe S, Mock J J, Smith D R and Padilla W J 2008 Perfect metamaterial absorber *Phys. Rev. Lett.* **100** 207402
[8] Li W and Valentine J 2014 Metamaterial perfect absorber based hot electron photo-detection *Nano Lett.* **14** 3510–4
[9] Hao J, Wang J, Liu X, Padilla W J, Zhou L and Qiu M 2010 High performance optical absorber based on a plasmonic metamaterial *Appl. Phys. Lett.* **96** 251104
[10] Pendry J B 2000 Negative refraction makes a perfect lens *Phys. Rev. Lett.* **85** 3966
[11] Fang N, Lee H, Sun C and Zhang X 2005 Sub-diffraction-limited optical imaging with a silver superlens *Science* **308** 534–7
[12] Schurig D, Mock J J, Justice 1 B, Cummer S A, Pendry J B, Starr A F and Smith D R 2006 Metamaterial electromagnetic cloak at microwave frequencies *Science* **314** 977–80
[13] Alù A and Engheta N 2005 Achieving transparency with plasmonic and metamaterial coatings *Phys. Rev.* E **72** 016623
[14] Cai W, Chettiar U K, Kildishev A V and Shalaev V M 2007 Optical cloaking with metamaterials *Nat. Photonics* **1** 224–7
[15] Modi A Y, Balanis C A, Birtcher C R and Shaman H N 2017 Novel design of ultrabroadband radar cross section reduction surfaces using artificial magnetic conductors *IEEE Trans. Antennas Propag.* **65** 5406–17
[16] Hawkes A M, Katko A R and Cummer S A 2013 A microwave metamaterial with integrated power harvesting functionality *Appl. Phys. Lett.* **103** 163901
[17] Ramahi O M, Almoneef T S, AlShareef M and Boybay M S 2012 Metamaterial particles for electromagnetic energy harvesting *Appl. Phys. Lett.* **101** 173903
[18] Chen Z, Guo B, Yang Y and Cheng C 2014 Metamaterials-based enhanced energy harvesting: a review *Phys.* B **438** 1–8
[19] Walia S, Shah C M, Gutruf P, Nili H, Chowdhury D R, Withayachumnankul W and Sriram S 2015 Flexible metasurfaces and metamaterials: a review of materials and fabrication processes at micro-and nanoscales *Appl. Phys. Rev.* **2** 011303
[20] Glybovski S B, Tretyakov S A, Belov P A, Kivshar Y S and Simovski C R 2016 Metasurfaces: from microwaves to visible *Phys. Rep.* **634** 1–72
[21] Holloway C L, Kuester E F, Gordon J A, O'Hara J, Booth J and Smith D R 2012 An overview of the theory and applications of metasurfaces: the two-dimensional equivalents of metamaterials *IEEE Antennas Propag. Mag.* **54** 10–35
[22] Chen H T, Taylor A J and Yu N 2016 A review of metasurfaces: physics and applications *Rep. Prog. Phys.* **79** 076401
[23] Minovich A E, Miroshnichenko A E, Bykov A Y, Murzina T V, Neshev D N and Kivshar Y S 2015 Functional and nonlinear optical metasurfaces *Laser Photonics Rev.* **9** 195–213
[24] Quarfoth R and Sievenpiper D 2013 Artificial tensor impedance surface waveguides *IEEE Trans. Antennas Propag.* **61** 3597–606
[25] Sheng P, Mei J, Liu Z and Wen W 2007 Dynamic mass density and acoustic metamaterials *Phys.* B **394** 256–61
[26] Chen S, Fan Y, Fu Q, Wu H, Jin Y, Zheng J and Zhang F 2018 A review of tunable acoustic metamaterials *Appl. Sci.* **8** 1480

[27] Mizukami K, Kawaguchi T, Ogi K and Koga Y 2021 Three-dimensional printing of locally resonant carbon-fiber composite metastructures for attenuation of broadband vibration *Compos. Struct.* **255** 112949
[28] Yu X, Zhou J, Liang H, Jiang Z and Wu L 2018 Mechanical metamaterials associated with stiffness, rigidity and compressibility: a brief review *Prog. Mater Sci.* **94** 114–73
[29] Meng Z, Liu M, Yan H, Genin G M and Chen C Q 2022 Deployable mechanical metamaterials with multistep programmable transformation *Sci. Adv.* **8** eabn5460
[30] Usta F, Scarpa F, Türkmen H S, Johnson P, Perriman A W and Chen Y 2021 Multiphase lattice metamaterials with enhanced mechanical performance *Smart Mater. Struct.* **30** 025014
[31] Haghpanah B, Salari-Sharif L, Pourrajab P, Hopkins J and Valdevit L 2016 Multistable shape-reconfigurable architected materials *Adv. Mater.* **28** 7915–20
[32] Ai L and Gao X L 2018 Three-dimensional metamaterials with a negative Poisson's ratio and a non-positive coefficient of thermal expansion *Int. J. Mech. Sci.* **135** 101–13
[33] Florijn B, Coulais C and van Hecke M 2014 Programmable mechanical metamaterials *Phys. Rev. Lett.* **113** 175503
[34] Wu L, Li B and Zhou J 2016 Isotropic negative thermal expansion metamaterials *ACS Appl. Mater. Interfaces* **8** 17721–7
[35] Chen P Y, Farhat M and Alù A 2011 Bistable and self-tunable negative-index metamaterial at optical frequencies *Phys. Rev. Lett.* **106** 105503
[36] Tao R, Xi L, Wu W, Li Y, Liao B, Liu L and Fang D 2020 4D printed multi-stable metamaterials with mechanically tunable performance *Compos. Struct.* **252** 112663
[37] Rao Y, Yan Y, Mei H, Zhou S, Tan Y, Cheng L and Zhang L 2022 3D-printed lattice structures with SiC whiskers to strengthen thermal metamaterials *Ceram. Int.* **48** 32283–9
[38] Wang G, Chen X, Liu S, Wong C and Chu S 2016 Mechanical chameleon through dynamic real-time plasmonic tuning *ACS Nano* **10** 1788–94
[39] Vicari A, Holman M and Schiavo A 2019 Metamaterials Market Forecast (Lux Research)
[40] Chen Y, Frenzel T, Guenneau S, Kadic M and Wegener M 2020 Mapping acoustical activity in 3D chiral mechanical metamaterials onto micropolar continuum elasticity *J. Mech. Phys. Solids* **137** 103877
[41] Gradoni G, Di Renzo M, Diaz-Rubio A, Tretyakov S, Caloz C, Peng Z and Phang S 2021 *Smart Radio Environments.* arXiv preprint arXiv:2111.08676.
[42] Molero C, Palomares-Caballero Á, Alex-Amor A, Parellada-Serrano I, Gamiz F, Padilla P and Valenzuela-Valdés J F 2021 Metamaterial-based reconfigurable intelligent surface: 3D meta-atoms controlled by graphene structures *IEEE Commun. Mag.* **59** 42–8
[43] Jimenez C M, Menargues E and García-Vigueras M 2020 All-metal 3-D frequency-selective surface with versatile dual-band polarization conversion *IEEE Trans. Antennas Propag.* **68** 5431–41
[44] Al Sabouni-Zawadzka A and Gilewski W 2019 Soft and stiff simplex tensegrity lattices as extreme smart metamaterials *Materials* **12** 187
[45] Reinbold J, Frenzel T, Münchinger A and Wegener M 2019 The rise of (chiral) 3D mechanical metamaterials *Materials* **12** 3527
[46] Wu W, Hu W, Qian G, Liao H, Xu X and Berto F 2019 Mechanical design and multifunctional applications of chiral mechanical metamaterials: a review *Mater. Des.* **180** 107950

[47] Gorshkov V, Sareh P, Navadeh N, Tereshchuk V and Fallah A S 2021 Multi-resonator metamaterials as multi-band metastructures *Mater. Des.* **202** 109522

[48] Georgakopoulos S V *et al* 2021 Origami antennas *IEEE Open J. Antennas Propag.* **2** 1020–43

[49] Wang H, Chen Q, Zetterstrom O and Quevedo-Teruel O 2021 Three-dimensional broad-band and isotropic double-mesh twin-wire media for meta-lenses *Appl. Sci.* **11** 7153

[50] Chen Z, Wang C F and Hoefer W J 2022 A unified view of computational electromagnetics *IEEE Trans. Microw. Theory Tech.* **70** 955–69

[51] Tan E L 2007 Unconditionally stable LOD–FDTD method for 3-D Maxwell's equations *IEEE Microw. Wirel. Compon. Lett.* **17** 85–7

[52] Elsherbeni A Z and Demir V 2015 *The Finite-Difference Time-Domain in Electromagnetics (No. 10764)* (Stevenage: SciTech Publishing)

[53] Yee K 1966 Numerical solution of initial boundary value problems involving Maxwell's equations in isotropic media *IEEE Trans. Antennas Propag.* **14** 302–7

[54] Aoki K, Guimard D, Nishioka M, Nomura M, Iwamoto S and Arakawa Y 2008 Coupling of quantum-dot light emission with a three-dimensional photonic-crystal nanocavity *Nat. Photonics* **2** 688–92

[55] Ho Y L D, Ivanov P S, Engin E, Nicol M F, Taverne M P, Hu C and Rarity J G 2011 FDTD simulation of inverse 3-D face-centered cubic photonic crystal cavities *IEEE J. Quantum Electron.* **47** 1480–92

[56] Zheng X, Taverne M P, Ho Y L D and Rarity J G 2018 Cavity design in woodpile based 3D photonic crystals *Appl. Sci.* **8** 1087

[57] Rumpf R C 2012 Simple implementation of arbitrarily shaped total-field/scattered-field regions in finite-difference frequency-domain *Prog. Electroma. Res.* B **36** 221–48

[58] Rumpf R C, Garcia C R, Berry E A and Barton J H 2014 Finite-difference frequency-domain algorithm for modeling electromagnetic scattering from general anisotropic objects *Prog. Electroma. Res.* B **61** 55–67

[59] Ivinskaya A M, Lavrinenko A V and Shyroki D M 2011 Modeling of nanophotonic resonators with the finite-difference frequency-domain method *IEEE Trans. Antennas Propag.* **59** 4155–61

[60] Zhai Y B, Ping X W, Jiang W X and Cui T J 2010 Finite-element analysis of three-dimensional axisymmetrical invisibility cloaks and other metamaterial devices *Commun. Comput. Phys.* **8** 823

[61] Burger S, Klose R, Schaedle A, Schmidt F and Zschiedrich L W 2005 FEM modeling of 3D photonic crystals and photonic crystal waveguides *In Integrated Optics: Devices, Materials, and Technologies IX* vol 5728 (Bellingham, WA: SPIE) 164–73

[62] Frenzel T, Hahn V, Ziemke P, Schneider J L G, Chen Y, Kiefer P and Wegener M 2021 Large characteristic lengths in 3D chiral elastic metamaterials *Commun. Mater.* **2** 4

[63] Alqurashi M M, Ganash E A and Altuwirqi R M 2022 Simulation of a low concentrator photovoltaic system using COMSOL *Appl. Sci.* **12** 3450

[64] Gibson W C 2021 *The Method of Moments in Electromagnetics* (Boca Raton, FL: CRC Press)

[65] Paez-Rueda C I, Fajardo A, Pérez M and Perilla G 2021 Closed-form expressions for numerical evaluation of self-impedance terms involved on wire antenna analysis by the method of moments *Electronics* **10** 1316

[66] Florencio R, Boix R R and Encinar J A 2018 Efficient spectral domain MoM for the design of circularly polarized reflectarray antennas made of split rings *IEEE Trans. Antennas Propag.* **67** 1760–71
[67] Florencio R, Somolinos Á, González I and Cátedra F 2020 Fast preconditioner computation for BICGSTAB-FFT method of moments with NURBS in large multilayer structures *Electronics* **9** 1938
[68] Silveirinha M G and Fernandes C A 2004 A hybrid method for the efficient calculation of the band structure of 3-D metallic crystals *IEEE Trans. Microw. Theory Tech.* **52** 889–902
[69] Tihon D, Sozio V, Ozdemir N A, Albani M and Craeye C 2016 Numerically stable eigenmode extraction in 3-D periodic metamaterials *IEEE Trans. Antennas Propag.* **64** 3068–79
[70] Tong M S, Qian Z G and Chew W C 2010 Nyström method solution of volume integral equations for electromagnetic scattering by 3D penetrable objects *IEEE Trans. Antennas Propag.* **58** 1645–52
[71] Chen D, Cho M H and Cai W 2018 Accurate and efficient Nystrom volume integral equation method for electromagnetic scattering of 3-D metamaterials in layered media *SIAM J. Sci. Comput.* **40** B259–82
[72] Engheta N, Murphy W D, Rokhlin V and Vassiliou M S 1992 The fast multipole method (FMM) for electromagnetic scattering problems *IEEE Trans. Antennas Propag.* **40** 634–41
[73] Fall M, Boutami S, Glière A, Stout B and Hazart J 2013 Multilevel fast multipole method based on a potential formulation for 3D electromagnetic scattering problems *J. Opt. Soc. Am.* A **30** 1273–80
[74] Albani M and Capolino F 2011 Wave dynamics by a plane wave on a half-space metamaterial made of plasmonic nanospheres: a discrete Wiener–Hopf formulation *JOSA* B **28** 2174–85
[75] Campione S and Capolino F 2012 Ewald method for 3D periodic dyadic Green's functions and complex modes in composite materials made of spherical particles under the dual dipole approximation *Radio Sci.* **47** 1–11
[76] Campione S and Capolino F 2016 Electromagnetic coupling and array packing induce exchange of dominance on complex modes in 3D periodic arrays of spheres with large permittivity *JOSA* B **33** 261–70
[77] Stevanoviæ I and Mosig J R 2007 Periodic Green's function for skewed 3-D lattices using the Ewald transformation *Microw. Opt. Technol. Lett.* **49** 1353–7
[78] Costa F, Monorchio A and Manara G 2014 An overview of equivalent circuit modeling techniques of frequency selective surfaces and metasurfaces *Appl. Comput. Electroma. Soc. J. (ACES)* **29** 960–76
[79] Molero Jiménez C 2021 Phase-resonance exploitation in full-metal 3D periodic structures for single-and multi-wideband applications *TechRxiv* https://doi.org/10.36227/techrxiv.17439956.v1
[80] Molero C and García-Vigueras M 2019 Circuit modeling of 3-D cells to design versatile full-metal polarizers *IEEE Trans. Microw.Theory Tech.* **67** 1357–69
[81] Algarín J M, Freire M J and Lapine M 2010 Ab initio experimental analysis of realistic resonant ring metamaterial lenses *In 2010 IEEE Antennas and Propagation Society International Symposium* (Piscataway, NJ: IEEE) pp 1–4
[82] Silveirinha M G 2009 Artificial plasma formed by connected metallic wires at infrared frequencies *Phys. Rev.* B **79** 035118

[83] Belov P A, Marques R, Maslovski S I, Nefedov I S, Silveirinha M, Simovski C R and Tretyakov S A 2003 Strong spatial dispersion in wire media in the very large wavelength limit *Phys. Rev.* B **67** 113103

[84] Sakhno D, Koreshin E and Belov P A 2021 Longitudinal electromagnetic waves with extremely short wavelength *Phys. Rev.* B **104** L100304

[85] Powell A W, Mitchell-Thomas R C, Zhang S, Cadman D A, Hibbins A P and Sambles J R 2021 Dark mode excitation in three-dimensional interlaced metallic meshes *ACS Photonics* **8** 841–6

[86] Saha S, Begam N, Biswas S and Sarkar P P 2021 A cascaded tunable wide stop band frequency selective surface with high roll-off band edge *Radioengineering* **30** 89–95

[87] Li H, Li B and Zhu L 2019 Wideband linear-to-circular polarizer based on orthogonally inserted slot-line structures *IEEE Antennas Wirel. Propag. Lett.* **18** 1169–73

[88] Álvarez H F, Cadman D A, Goulas A, de Cos Gómez M E, Engstrøm D S, Vardaxoglou J C and Zhang S 2021 3D conformal bandpass millimeter-wave frequency selective surface with improved fields of view *Sci. Rep.* **11** 12846

[89] Wang X, Cheng Y and Dong Y 2021 Millimeter-wave dual-polarized metal transmitarray antenna with wide gain bandwidth *IEEE Antennas Wirel. Propag. Lett.* **21** 381–5

[90] Li B, Mei C Y, Zhou Y and Lv X 2020 A 3-D-printed wideband circularly polarized dielectric reflectarray of cross-shaped element *IEEE Antennas Wirel. Propag. Lett.* **19** 1734–8

[91] Yang Y, Wang W, Moitra P, Kravchenko I I, Briggs D P and Valentine J 2014 Dielectric meta-reflectarray for broadband linear polarization conversion and optical vortex generation *Nano Lett.* **14** 1394–9

[92] Cui Y, Nauroze S A, Bahr R and Tentzeris E M 2020 3D Printed one-shot deployable flexible 'Kirigami' dielectric reflectarray antenna for mm-wave applications *2020 IEEE/MTT-S Int. Microwave Symp. (IMS)* (Piscataway: IEEE) 1164–7

[93] Wu G B, Zeng Y S, Chan K F, Qu S W and Chan C H 2019 3-D printed circularly polarized modified Fresnel lens operating at terahertz frequencies *IEEE Trans. Antennas Propag.* **67** 4429–37

[94] Omar A A and Shen Z 2016 Multiband high-order bandstop 3-D frequency-selective structures *IEEE Trans. Antennas Propag.* **64** 2217–26

[95] Celozzi S, Araneo R, Burghignoli P and Lovat G 2023 Frequency selective surfaces *Electromagnetic Shielding: Theory and Applications* (New York: Wiley) 363–408

[96] Zhu J, Hao Z, Wang C, Yu Z, Huang C and Tang W 2019 Dual-band 3-D frequency selective surface with multiple transmission zeros *IEEE Antennas Wirel. Propag. Lett.* **18** 596–600

[97] Pelletti C, Bianconi G, Mittra R and Shen Z 2013 Frequency selective surface with wideband quasi-elliptic bandpass response *Electron. Lett.* **49** 1052–3

[98] Ehrenberg I M, Sarma S E and Wu B I 2012 A three-dimensional self-supporting low loss microwave lens with a negative refractive index *J. Appl. Phys.* **112** 073114

[99] Zhu D Z, Gregory M D, Werner P L and Werner D H 2018 Fabrication and characterization of multiband polarization independent 3-D-printed frequency selective structures with ultrawide fields of view *IEEE Trans. Antennas Propag.* **66** 6096–105

[100] Harnois M, Himdi M, Yong W Y, Rahim S K A, Tekkouk K and Cheval N 2020 An improved fabrication technique for the 3-D frequency selective surface based on water transfer printing technology *Sci. Rep.* **10** 1714

[101] Nauroze S A, Novelino L S, Tentzeris M M and Paulino G H 2018 Continuous-range tunable multilayer frequency-selective surfaces using origami and inkjet printing *Proc. Natl Acad. Sci.* **115** 13210–5
[102] Burckel D B, Wendt J R, Ten Eyck G A, Ellis A R, Brener I and Sinclair M B 2010 Fabrication of 3D metamaterial resonators using self-aligned membrane projection lithography *Adv. Mater.* **22** 3171–5
[103] Xiong X, Xue Z H, Meng C, Jiang S C, Hu Y H, Peng R W and Wang M 2013 Polarization-dependent perfect absorbers/reflectors based on a three-dimensional metamaterial *Phys. Rev.* B **88** 115105
[104] Matthaiou M, Yurduseven O, Ngo H Q, Morales-Jimenez D, Cotton S L and Fusco V F 2021 The road to 6G: ten physical layer challenges for communications engineers *IEEE Commun. Mag.* **59** 64–9
[105] Chen R, Long W X, Wang X and Jiandong L 2020 Multi-mode OAM radio waves: generation, angle of arrival estimation and reception with UCAs *IEEE Trans. Wirel. Commun.* **19** 6932–47

www.ingramcontent.com/pod-product-compliance
Ingram Content Group UK Ltd.
Pitfield, Milton Keynes, MK11 3LW, UK
UKHW051851100926
8556IPUK00005B/576

9 780750 354202